Herschel Clifford Parker

A Systematic Treatise on Electrical Measurements

Herschel Clifford Parker

A Systematic Treatise on Electrical Measurements

ISBN/EAN: 9783337036119

Printed in Europe, USA, Canada, Australia, Japan

Cover: Foto ©berggeist007 / pixelio.de

More available books at **www.hansebooks.com**

A Systematic Treatise

ON

ELECTRICAL

MEASUREMENTS

BY

HERSCHEL C. PARKER, Ph. B.,

Tutor in Physics, Columbia University. Instructor in Electrical Measurements
Associate Member of the American Institute of Electrical Engineers.

NEW YORK:
SPON & CHAMBERLAIN, 12 CORTLANDT ST.
LONDON:
E. & F. N. SPON, LIMITED. 125 STRAND.
1897.

Press of McIlroy & Emmet. 36 Cortlandt St., N. Y.

NOTE.

The present Treatise on Electrical Measurements recently appeared as a series of articles in an electrical monthly and has been bound in book form with but very little revision. It should, therefore, not be judged as a finished work.

The method of classification here made use of has been found very satisfactory in several courses of lectures given by the writer to students in Electrical Engineering at Columbia University.

CONTENTS.

INTRODUCTION.

AN ACCURATE knowledge of electrical measurement, to the electrical engineer as well as the physicist, is of the first importance.

It is a branch of science where engineering and physics meet.

There appears, however, in many instances, to be a lack of uniformity in the methods employed. Indeed, it almost seems as if there were two schools of electrical measurement. But this is probably due, to some extent at least, to the lack of a proper co-ordination and classification of the subject.

New methods of practice have rapidly developed, and improved instruments are constantly coming into use, so that it is not strange if there be a little confusion.

Thus we may find a text-book that is almost perfect as far as resistance work is concerned, but deficient with regard to E.M.F. or current, describing at length obsolete methods and entirely omitting many of the best ones. So that often the student may be compelled to consult a great number of standard works and supplement this by long personal observation to obtain even a fair comprehension of the practical methods.

The subject, it seems to the writer, should be attacked in the most systematic manner and the classification thoroughly worked out. Indeed, classification and knowledge are very nearly synonymous terms.

What follows is offered as an example of such a method of treatment.

It seemed advisable to make the classification fairly complete, and then to clearly point out the most desirable methods or those applicable to any particular case.

Of course, there are many omissions and possibly errors ; but it is hoped that it will facilitate the acquirement of a working knowledge of the subject by students of electrical engineering.

In the text of the treatise, free use has been made of the standard works, especially "Kempe's Hand-book of Electrical Testing," but it is also believed that a considerable amount of new material is presented.

CHAPTER I.

CLASSIFICATION OF ELECTRICAL MEASUREMENTS.

RESISTANCE.

LOW RESISTANCE.
1. Thomson's Double Bridge.*
2. Differential Galvanometer.*
3. Projection of Potentials.
4. Fall of Potential.
5. Potentiometer.
6. Carey Foster's Method.

MEDIUM RESISTANCE.

WHEATSTONE BRIDGE.

Wire Bridge (Variable Ratio)
- Straight (Metre).
- Circular (Kohlrausch).
- Parallel (Poggendorff).
- Direct Reading (Kirchhoff).

Slide Coil (Variable Ratio.)
- Five Arc (Cushman).*
- Quadruplex (Muirhead)
- Duplex (Varley).

Constant Ratio
- Post Office Bridge.
- Decade Bridge.*

"SPECIFIC RESISTANCE."
- P. O. Bridge.*
- Conductivity Balance.

COMPARISON OF STANDARDS.
- Carey Foster's Method.*
- Substitution in the Bridge.

CALIBRATION.

Bridge Wire
- Comparison with Rheostat (Potentiometer Method.)*
- Carey Foster's Method.
- Double bridge.
- Differential Galvanometer.

Rheostat
- Substitution in the Bridge.

GALVANOMETER RESISTANCE.
- P. O. Bridge.*
- Thomson's Method.
- ½ Deflection.

HIGH RESISTANCE.
- Slide Coil Bridge.*
- Potentiometer Method.
- Deflection Method.*
- Loss of Charge.

3

INSULATION........

1. *Insulite.*—" Specific Insulation."

2. *Insulated Wires.*
   - Short Lengths.
   - Cable. { Single Core. / Multiple Core.
   - Joint Testing. { Loss of charge.* / Accumulation. / Electrometer.

3. *Aerial Wires.*

RESISTANCE OF TELEGRAPH LINES, CABLES, ETC.

1. *P. O. Bridge.*
2. *Loop Test.....* { *a.* / *b.*
3. *Equilibrium.*
4. *Mance's Method.*
5. *Equal Deflection.*

LOCALIZATION OF FAULTS.

1. *Complete Fault in Insulation.*
2. *Partial* " " " (Earth Resistance.)
3. *Variable* " " " (Polarization or Cable Current.)
4. *Fault plus E. M. F.* (Earth Current.)
5. *Fault in Conductor.*
6. *Faults of High Resistance.*

BATTERY RESISTANCE.

1. *Fall of Potential.*.......... { Condenser. / High Resistance. / Voltmeter.*
2. *Added Resistance...........* { Tangent Galvanometer. / ·½ Deflection.
3. *Mance's Method.*
4. *Current and E. M. F.**

RESISTANCE OF ELECTROLYTES. { *Constant Current.* / *Alternating Current.**

INCANDESCENT LAMPS, " DYNAMO RESISTANCE," ETC. { *Fall of Potential.* / *Current and E. M. F.* · / *Ohmmeter.*

DETERMINATION OF THE OHM.

ELECTROMOTIVE FORCE.

BATTERIES AND DIRECT CURRENTS.

*High Resistance Method.**......... { Deflection and Resistance. / Equal Deflection. / Equal Resistance.

Wheatstone's Method.
Lumsden's "
Condenser " *

*Potentiometer.**... { Five Arc (Cushman.) / Quadruplex (Muirhead.) / Duplex (Varley.)

Current and Resistance.
Electrometer.
*Voltmeter.**

ALTERNATING CURRENTS.

Electrometer ....
- Quadrant.
- Multicellular.*
- Electrostatic Voltmeter.* { Thomson's. Weston's.
- Low Reading.

Dynamometer ...
- Siemens'
- Weston's* (Alternating Current Voltmeter.)

Caloric Voltmeter (Cardew's.)*

Attraction Voltmeters...........
- Evershed's.
- Magnetic Vane, etc.

VERY HIGH E. M. F.
- Electrostatic Voltmeter.
- Absolute Electrometer.
- Striking Distance of Spark.

VERY LOW E. M. F.
- Galvanometer.*
- Voltmeter.*
- Capillary Electrometer (Lippmann's.)

CALIBRATION OF VOLTMETERS. Potentiometer.

STANDARDS OF E.M.F. Checking by Current and Resistance.

CURRENT.

DIRECT CURRENTS.

E. M. F. and Resistance.........
- Direct Method.
- Differential Method (Cardew's.)
- Bridge Method (Kempe's.)

P. D. and Resistance.........
- Direct Deflection Method.
- Equilibrium Method.
- Potentiometer "
- Voltmeter " *
- Galvanometer " *

Tangent Galvanometer.

Voltameters...... { Weight. Volume.

Ammeter.*

ALTERNATING CURRENTS.
- Dynamometer.
- Current Balance (Thomson's.)
- Attraction Ammeters........... { Evershed's, Schuckert's, etc.
- Calorimetric Methods.

CALIBRATION OF AMMETERS.

ABSOLUTE DETERMINATION (Tangent Galvanometer.)

ENERGY. .... $\left\{ \begin{array}{l} Wattmeter \ \ldots\ldots \ \left\{ \begin{array}{l} Weston's. \\ Siemens'. \end{array} \right. \\ Voltmeter\ and\ Ammeter. \end{array} \right.$

QUANTITY.......... $\left\{ \begin{array}{l} Voltameter. \\ `` Meters." \\ Ballistic\ Galvanometer. \end{array} \right.$

CAPACITY.

DEFLECTION
   METHODS. $\left\{ \begin{array}{l} Direct\ Deflection.* \\ Divided\ Charge. \\ \\ Loss\ of\ Charge\ldots \left\{ \begin{array}{l} Discharge. \\ Deflection. \end{array} \right. \end{array} \right.$

ZERO METHODS.... $\left\{ \begin{array}{l} Bridge\ Method. \\ Potentiometer\ Method \text{ (Mixtures.)} \end{array} \right.$

ABSOLUTE DETERMINATION (Ballistic Galvanometer.)

INDUCTANCE.

BRIDGE METHOD (Maxwell's.)

SECOHMMETER
   METHOD. $\left\{ \begin{array}{l} With\ Standard. \\ Without\ `` \end{array} \right.$

CONDENSER
   METHOD. $\left\{ \begin{array}{l} Deflection. \\ Zero. \end{array} \right.$

CALCULATION.

[IMPEDANCE.]

EFFICIENCY... .... $\left\{ \begin{array}{l} Cells. \\ Lamps. \\ Motors. \\ Transformers. \\ Dynamos. \end{array} \right.$

MAGNETIC
   DETERMINATIONS. $\left\{ \begin{array}{l} Field\ (\mathfrak{H}) \\ Intensity\ of\ Magnetization\ (\mathfrak{J}) \\ Permeability\ \left(\mu = \dfrac{\mathfrak{B}}{\mathfrak{H}}\right) \\ Susceptibility\ \left(\dfrac{\mathfrak{J}}{\mathfrak{H}}\right) \\ Hysteresis. \\ Magneto\text{-}Motive\ Force. \\ Reluctance. \end{array} \right.$

REMARKS.

In the above classification it is not attempted to give the various methods in the order of their relative merit, but rather according to their logical sequence.

What are believed to be the superior methods, or those especially applicable in any given case, are indicated by stars.

In the classification of resistance measurements it was thought advisable to give the different forms of the Wheatstone Bridge. It also seemed best to classify the different cases of " Insulation" and those that might occur in the "Localization of Faults."

The determination of " Energy " and of " Quantity " so closely approximates the measurement of current that they appear to belong as sub-headings under that subject.

" Impedance " is given as a special application of the measurement of " Inductance."

Under " Efficiency " are given several special cases.

To complete the subject, a number of Magnetic Determinations are added, for there certainly should be no fixed line drawn between electrical and magnetic measurements, considering the present state of electrical science.

CHAPTER II.

GALVANOMETERS.

MOVABLE MAGNET-IC SYSTEM.
{ *Tangent.*
Astatic.
Thomson......... { Single Coil.
Duplex.
Quadruplex.
Aperiodic.
Differential.
Ballistic.

MOVABLE COIL.....
{ *D'Arsonval.......* { Aperiodic.
Ballistic.

Weston Pattern (Portable)

Ayrton &
Mather Pattern.. { Aperiodic.
Ballistic.
Rowland "
Electro-Magnet Pattern.

GENERAL.

Figure of Merit. —By the " Figure of Merit " is meant the strength of current required to produce a deflection of one scale division, or the resistance that must be introduced into the circuit to reduce the deflection to one scale division with a P. D. of one volt.

All the conditions should be specified, such as the distance of the scale from the galvanometer, width of scale divisions * and time of vibration of galvanometer.

Example : { Deflection = 250 scale divisions ($\frac{1}{1000}$ shunt.)
P. D. = 2.5 volts.
Resistance = 100,000 ohms.

Figure of Merit = $\dfrac{2.5}{100,000 \times 250 \times 1,000}$ = 1×10^{-10} amperes.

or 1×10^{10} ohms, that is .0001 micro-amperes, or 10,000 meg-ohms.

Sensitiveness.—This term may be employed to indicate the P. D. across the galvanometer terminals necessary to give a deflection of one scale division. Since $E = C\,R$, it may be obtained by

* Unless otherwise specified it is assumed that a mm. scale is used at the distance of a metre from the galvanometer mirror.

multiplying the figure of merit by the resistance of the galvanometer.

Thus, if the galvanometer in the above example had a resistance of 10,000 ohms, its "Sensitiveness" would be .0001 micro-
ampere $\times$ 10,000 = 1 micro-volt.

In the measurement of low resistance it is, of course, desirable
to have a galvanometer of the maximum "efficiency." Usually,
galvanometers of low resistance have a greater efficiency than
those of high resistance, but the figure of merit increases with
the number of turns, and consequently high resistance galvanometers have a greater figure of merit. In the measurement of
high resistance and insulation the galvanometer resistance has
but little effect on the current, and hence it is best to employ a
galvanometer with the maximum figure of merit.

Shunts.—In order to vary the sensitiveness of galvanometers,
a portion of the current is deflected by a resistance in parallel
with the galvanometer. The value of this shunt is given by the
formula :

$$S = G \times \frac{1}{n-1,}$$

where S = resistance of shunt, G = resistance of galvanometer,
and $\frac{1}{n}$ = the portion of the current received by the galvanometer, (that is, the amount the galvanometer deflection is reduced
to, where such deflection is proportional to the current.)

$$\text{Example : } 111.1 = 1000 \times \frac{1}{10-1}$$

In this case the resistance of a one-tenth shunt for a 1000 ohm
galvanometer must be 111.1 ohms.

When the resistance in the main circuit is comparatively low,
the use of a shunt reduces the resistance of the entire circuit an
appreciable amount and introduces a certain error in the measurement unless a compensating resistance is added.

The formula

$$S^1 = G \times \frac{n-1}{n}$$

gives the value of this resistance. In the above example it
would be 900 ohms $\left(900 = 1,000 \times \frac{10-1}{10} \right)$

Errors.—With a reflecting galvanometer and a tangent scale,
the beam of light is deflected through twice the angle that the
mirror is turned. In an observation where two deflections are
taken, the error in assuming that the ratio of the tangents of
twice the angles is the same as the ratio of the tangents of the
angles may amount to one per cent. or over where the ratio
is greater than six to one.

The formula for the induction is :

$$c_1 : c_2 :: d_1 \left(\sqrt{l^2 + d_1^2} - l\right) : d_2 \left(\sqrt{l^2 + d_2^2} - l\right),$$

where d_1, d_2, represent the two deflections; c_1, c_2, the ratios of the two currents, and l, the distance from the scale to the mirror in scale divisions.

The error of observation where two deflections are taken, is increased the more widely the deflections differ. The formula is:

$$T = \frac{\dfrac{1}{m} \times 100}{d} \times (1 + n)$$

$T =$ percentage error of the determination, $\dfrac{1}{m} =$ error of observation in scale divisions, $d =$ first deflection, $n =$ ratio of first deflection to the second. Example :

$$T = \frac{\frac{1}{2} \times 100}{250} \times (1 + 4) = 1\%.$$

From the above considerations and also on account of possible variations in the E. M. F. during experiment, it is apparent that zero methods are preferable. When deflection methods are employed it is often practicable to so vary the *P. D.* that the two deflections may be nearly equal. The ratio of the differences of potential is then used in the calculation.

Angle of maximum sensitiveness. Where the strength of current and consequently the deflections are proportional to the tangents of the angles, the galvanometer has the greatest sensitiveness when the needle makes an angle of 45° with the coil. With a reflecting galvanometer, the angle of maximum sensitiveness is the largest angle that can be obtained, since the angle of deflection is but a very few degrees and, therefore, the true maximum angle can never be obtained.

TANGENT GALVANOMETER.

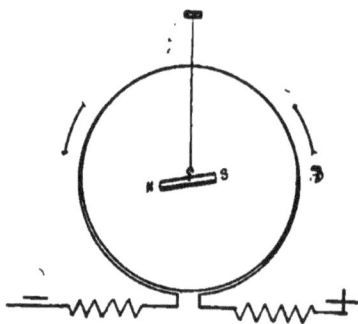

FIG. I.

In this form of galvanometer, a short magnetic needle is centrally placed within a coil of wire of large radius as shown in Fig. 1. The needle may carry a pointer moving over a graduated circle, or the deflections can be read by means of a mirror and telescope and scale, or lamp and scale.

If the influence of the coil

on the needle is the same whatever angle the needle makes with it, then the strength of the current circulating in the coil is directly proportional to the tangent of the angle of deflection.

Theoretically, to obtain this result, the magnet should be a mere point, but practically it is sufficient for the coil to be about ten times as large in diameter as the length of the needle.

This instrument furnishes a most convenient means for the comparison of current strengths, and is of the greatest interest on account of its employment in the absolute measurement of current and consequently in the absolute determination of the ohm.

The perfection of the ammeter, however, and the accuracy and facility with which current may be measured by determining the P. D. across a shunt, render the tangent galvanometer of far less importance than formerly to the electrical engineer.

Astatic Galvanometer.

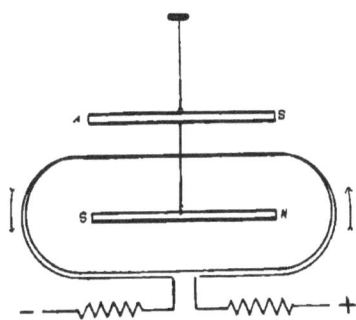

FIG. 2.

This instrument is of very simple construction and is quite sensitive. It is especially adapted for use with zero methods or may be employed as a sine galvanometer.

It consists of an astatic pair of needles of any convenient length suspended by a fibre arranged as in Fig. 2; one needle turns within the coil while the other moves above it.

If the coil is made to rotate and the angle of rotation measured, then when a current is sent through the galvanometer if the coil be turned until it is parallel with the needles, that is, if the needles are again brought to zero, the current is proportional to the sine of the angle of rotation. This is independent of the size or shape of the coil or the length of the needles.

Thomson Galvanometer.

The limits in the range of electrical measurement are usually fixed by the sensitiveness of the galvanometer.

The highest figure of merit, and perhaps the greatest efficiency may be obtained with the Thomson galvanometer.

The magnetic system is astatic and is formed of a number of small light magnets, usually pieces of watch-spring.

One set of these magnets is attached to the back of a small mirror, while another set is fixed to a light aluminium vane (Fig. 3)

The upper set of magnets moves in the centre of a large coil of wire of many turns, while the lower set is beneath the coil. This coil consists of two portions, and is hinged so that it may easily be opened and the magnetic system put in place. The high resistance galvanometers are usually furnished with two coils, and the more recent instruments with four coils. There is, of course, a set of magnets for each coil, and by this means the magnetic moment and number of turns may be greatly multiplied. (Fig. 4.) The entire magnetic system is suspended by means of the very finest fibre, either of silk or quartz.

The galvanometer is also

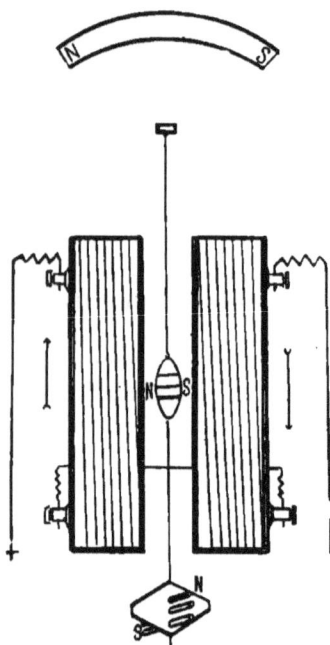

FIG. 3.

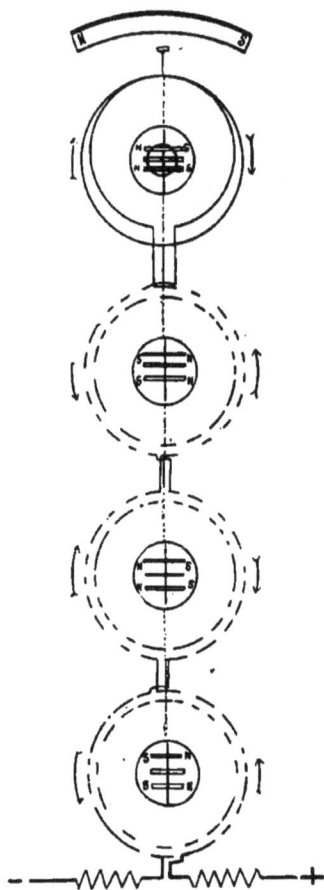

FIG. 4.

provided with a field magnet. This magnet is used to neutralize the earth's field, and, being nearer the upper set of magnets than the others, it also acts as a directing or controlling magnet.

The needles being very small, and each set being placed in

the axis of a large coil of wire which completely surrounds it, the tangents of the deflections are approximately directly proportional to the strength of the currents producing them.

Since the deflections are read by means of a reflected beam of light, the angle through which this beam of light turns will be twice the angle through which the mirror turns, and, consequently, the deflections will be proportional to the tangents of twice the angles.

If, however, the deflections to be compared are both small, or if they do not differ greatly, these deflections may be taken as proportional to currents producing them.

Considerable care is required in setting up the galvanometer.

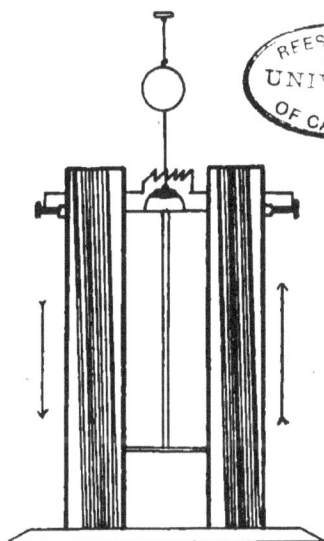

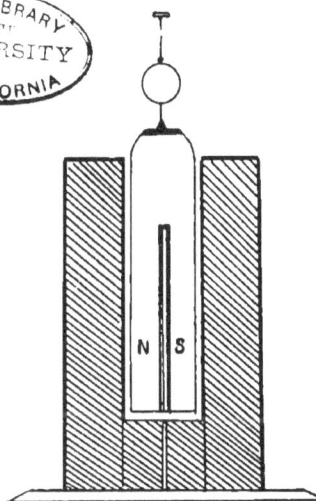

FIG. 5.   FIG. 6.

A position should be selected as free from vibrations and magnetic disturbances as possible, and all torsion should be carefully removed from the fibre.

The magnetic system is usually not quite astatic, and if the field magnet is removed, will set in the magnetic meridian. The field magnet should then be lowered until it just neutralizes the earth's field. This is shown by the magnetic system being in unstable equilibrium. The control magnet is then raised slightly above this position.

Two control magnets are sometimes employed to render the field more uniform.

The scale should be placed parallel to the mirror.

The high resistance galvanometers are wound with a resistance varying from 3,500 ohms to 100,000 ohms, and as great a figure of merit as 100,000 meg-ohms may be obtained. This form of instrument is extremely useful in insulation work.

The low resistance galvanometers are extraordinarily sensitive but are also very easily affected by thermal currents, so that it is sometimes found advisable to use some form of D'Arsonval galvanometer for low resistance determinations.

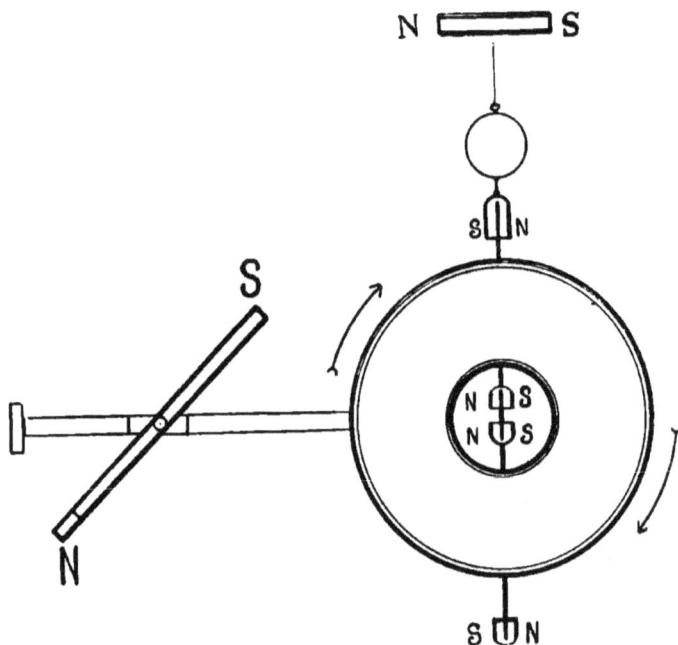

FIG. 7.

Aperiodic Galvanometer.

In this galvanometer a large bell magnet is employed. This magnet is enclosed between massive plates of copper, and hence when set in motion it induces powerful Foucault currents in the copper which quickly bring the magnet to rest. ·(Figs. 5 and 6.)

Moreover, slight changes in the surrounding magnetic field seem to produce but little effect on this form of galvanometer, and a rather high figure of merit may easily be obtained even where no great care has been exercised in its construction.

DIFFERENTIAL GALVANOMETER.

A galvanometer may be made differential by winding two wires on the same coil or by the use of two coils, and then sending the current through in opposite directions. It is adjusted by sending the same current through each coil and adding resistance to one of the coils until no deflection is produced. Where two coils are employed, a rough adjustment may be made by moving one of the coils. It is always better, however, to make the final adjustment by the addition of resistance.

The coils of a Thomson galvanometer may be connected differentially.

The differential galvanometer is useful in the measurement of low resistance.

BALLISTIC GALVANOMETER.

For certain determinations a galvanometer with a long period of vibration, a large moment of inertia, and with but very little decrement is required.

One of the standard forms of this instrument is shown in the diagram. (Fig. 7.)

Four bell magnets are employed, a coil of high resistance, a control magnet and a directing magnet. It is desired to give the moving magnetic system considerable weight and yet have the air resistance as little as possible.

The mirror used should be very small, and the suspending fibre extremely fine and without torsion.

This form of galvanometer is exceedingly sensitive to vibrations and changes in the magnetic field. The deflections are controlled by means of a check coil placed near the instrument.

THE D'ARSONVAL GALVANOMETER.

The radical difference between this form of galvanometer and those previously described, is that here the coil is movable and the magnets are fixed as shown in Fig. 8. A very intense field is obtained by the combination of several horse-shoe magnets having a soft iron core or a compound magnet placed between their poles.

In this space moves the rectangular coil wound on a thin copper or silver frame. The Foucault currents induced in this frame render the galvanometer almost aperiodic. The resisting force is the torsion of the suspending wires. Consequently, assuming the field to be uniform, the currents flowing through the coil are directly proportional to the angular deflection.

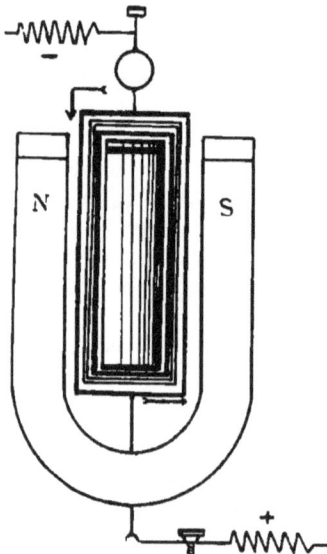

FIG. 8.

Hence, for accurate work in deflection methods, a circular scale should be employed or the tangential deflections reduced to the corresponding angles. ·

On account of the great strength of field, this galvanometer is scarcely effected by considerable magnetic changes, even in its immediate vicinity. ·

The coil is usually wound to about 100 ohms, but a higher resistance may be used when desired. The figure of merit that may be obtained, expressed in megohms, is usually somewhat less than the galvanometer resistance in ohms.

For nearly the whole range of electrical measurement, this galvanometer is probably the most satisfactory one that can be employed.

When the metallic frame is not used, the decrement becomes very small ; and since the weight of the coil is considerable, the moment of inertia is large. That is, the conditions of a ballistic galvanometer are fulfilled. The moment of inertia may be still further increased by adding a weight to the coil. The coil may be brought to rest by means of a short circuiting key.

Weston Pattern.—If the coil, instead of being suspended by a wire turns in jeweled bearings, the current being lead in by delicate watch-springs (Fig. 9) ; and if the horse-shoe magnet or magnets be placed horizontally. the coil having an oblique position between the poles, the gal-

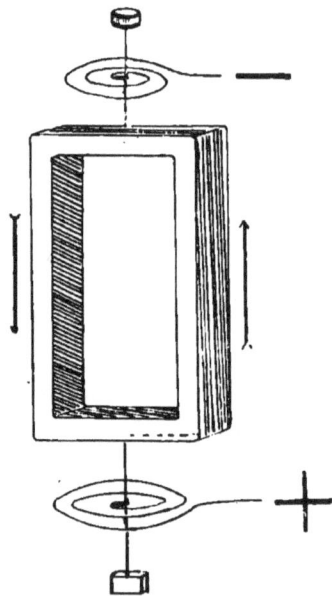

FIG. 9.

vanometer is then of the Weston pattern. This form of instrument is generally used in combination with a low shunt resistance or a high series resistance and then constitutes the well-known Weston ammeter or Weston voltmeter. Without these auxiliary resistances, however, it is one of the very best forms of D'Arsonval galvanometer. The sensitiveness is almost as great, while it is far more compact and portable than the ordinary

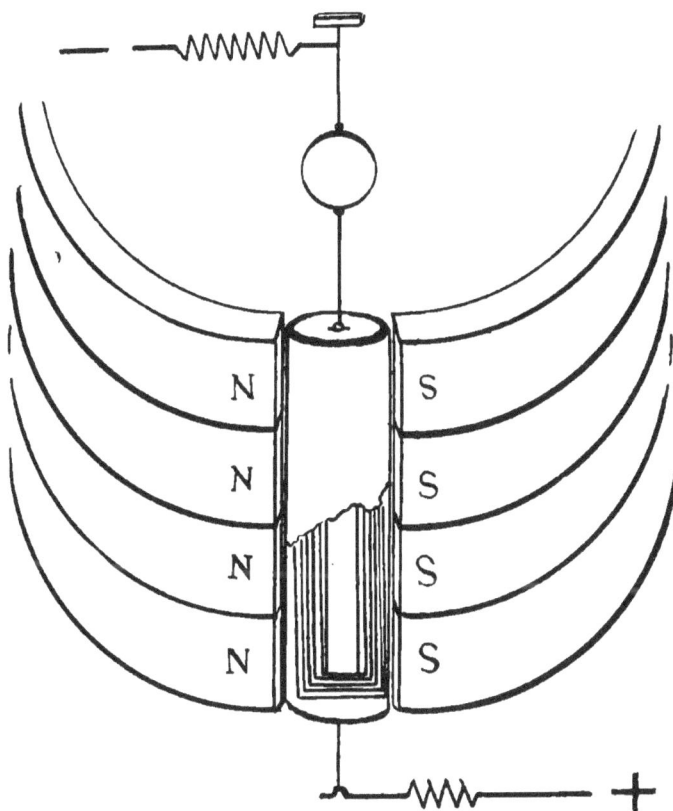

FIG. 10.

galvanometer. Besides this, the angular deflections are accurately proportional to the currents producing them.

Ayrton and Mather.—In this form of galvanometer (Fig. 10) a large number of circular magnets are placed horizontally, the poles being brought very near together. In this small space is suspended a long narrow coil. The coil carries a light metallic sheath where aperiodicity is desired, but is without the sheath when used ballistically.

The coil may be wound to as high a resistance as 4,000 ohms, and a figure of merit of over 1,000 meg-ohms may be obtained.

This is probably the best form of ballistic galvanometer and may also be employed in ordinary insulation work.

Rowland.—The Rowland D'Arsonval galvanometer has an elliptically shaped permanent magnet enclosed between two faces of sheet brass, thus forming a closed space in which the coil swings. The coil is provided with a large mica vane for dampening its vibrations.

The pole faces of the magnet are so shaped that the deflections, as read with a telescope or lamp and scale, are said to be exactly proportional to the current passing.

The coil may be given a resistance of 1,500 ohms and a figure of merit of 500 meg-ohms obtained.

Electro-Magnet Pattern.—The strength of field, and consequently the sensitiveness of the galvanometer may be increased by the use of electro-magnets. This, however, is only necessary in special cases of research work.

Conclusions.—From the above considerations it is evident that the Thomson galvanometer, having the highest figure of merit and the greatest sensitiveness is the most desirable for either very high or very low resistance determinations. In certain cases of low resistance work, however, on account of thermal currents, a D'Arsonval galvanometer may prove preferable. For all ordinary measurements a D'Arsonval galvanometer is recommended. For the determination of high insulation it is best to employ a Thomson high resistance galvanometer.

CHAPTER III.

Low Resistance.

Low resistance is, of course, a relative term ; but it is here used to indicate resistances too small to be accurately determined by the ordinary Wheatstone bridge methods. For most resistance measurements, an accuracy of about one per cent. is desirable. That is, if .01 ohm is to be measured, it requires the determination to be made to .0001 ohm. But contact resistances are an unknown variable, and may easily introduce an error as great as the last named figure.

Since the effect of these contact resistances can never be entirely eliminated with the Wheatstone Bridge, the upper limit of low resistance may be placed at about .01 ohm.

It is true that there is a form of Wheatstone Bridge that reads to .000001 ohm, but measuring to a millionth of an ohm and reading to a millionth of an ohm are very different affairs.

In most of the special methods described, the effect of contact resistance is practically eliminated.

It should also be understood that when a measurement is made to, say, 1×10^{-7} ohm, the entire resistance is not much more than 1×10^{-4} ohm. For it is hardly possible in low resistance work to measure better than .1 per cent. under the most favorable conditions, on account of temperature coefficients and thermal effects.

The measurement of such extremely low resistances is made possible by Ohm's law, $C = \dfrac{E}{R}$.

One of the limiting conditions is the sensitiveness of the galvanometer, and the galvanometer deflections are approximately proportional to the current. The current, however, is inversely proportional to the resistance.

Hence, the lower the resistance, the greater the current for any given P. D. That is, the smaller the resistance in circuit, the lower the limit of the measurement becomes.

But, of course, there is a constant and limiting resistance due to the battery, conducting wires, etc., no matter how small the resistance to be measured is.

The case very roughly approximates that of a balance whose sensitiveness varies inversely as the weights to be determined.

Thomson's Double Bridge.—This is one of the most satisfactory and convenient methods for the measurement of low resistance.

The arrangement of the experiment is shown by the diagram, Fig. 11.

If the ratio coils A, A', are each made equal to 100 ohms, and the coils B, B', each equal to 10 ohms, then when R : x :: 100 : 10

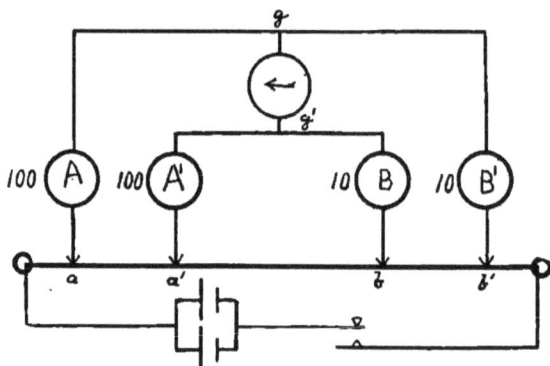

FIG. 11.

there will be no deflection of the galvanometer on closing the circuit.

R is the known variable resistance and x is the resistance to be determined.

The principle is as follows : If there be no junction between R and x, that is, if $r = \infty$, and then if R and x are zero, we have the ordinary case of the Wheatstone Bridge, 100 : 10 :: 100 : 10, and the potential at g, g', will be the same.

If x is given a value, and R is changed until equal to 10 x, we have:

$$100 : 10 :: 100 + 10\,x : 10 + x, \text{ or } \frac{100}{10} = \frac{10\,(10 + x)}{10 + x} = \frac{10}{1},$$

which is again the condition of equal potential at g and g'. That is, 100 : 10 :: R : x, when the galvanometer shows no deflection

If now R and X be joined by any resistance r, if R be zero, we have :

$$\frac{R}{X} = \frac{100}{10} = \frac{10}{1}, \text{ or } 100 : 10 :: R : X.$$

But zero and $\propto$ are the limiting values for small R; hence whatever the resistance of r, it does not effect the potential at g, g'. That is, it simply shunts off a portion of the current, leaving the potential at g, g' still the same.

The resistance of the movable contacts a, a', and b, b', together with the wires leading to them, is added to that of the ratio coils. Hence the resistance of these coils should be so great that the above mentioned resistance may be negligible compared to them. The smaller coils ought not to be less than 10 ohms.

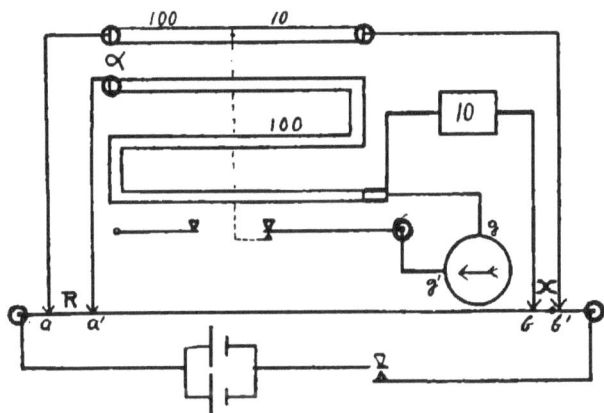

FIG. 12.

The contact points a, a', and b, b', must occupy the same relative positions as shown in the diagram. If, for instance, they should be placed in the positions a, a', b', b, it would be impossible to obtain a balance.

Again, on first setting up the experiment, the terminals a, a', should be joined to one point in the circuit, and the terminals b, b', to some other point in the circuit. Then if, on closing the key, there is any deflection, a compensating resistance should be added to one of the coils until there is no deflection.

Special coils may be used for the ratios, or a P. O. bridge and rheostat can be employed according to Fig. 12.

A portable form of Thomson's bridge is manufactured by Siemens & Halske. The standard low resistance is a thick wire, stretched around the instrument, and a movable contact is arranged so as to include more or less of the wire. Peg resistances are arranged so that the resistance can be multiplied or divided, so that the range of the instrument is very large.

A good method of procedure is the following : A metre of G. S. wire of about 0.1 ohm resistance is accurately measured on the P. O. bridge, and its resistance per mm. calculated. The wire is then stretched over a metre stick, and used as the known resistance in the double bridge to determine the resistance of a second metre of copper wire. This last wire is employed to measure the resistance of a still larger copper wire.

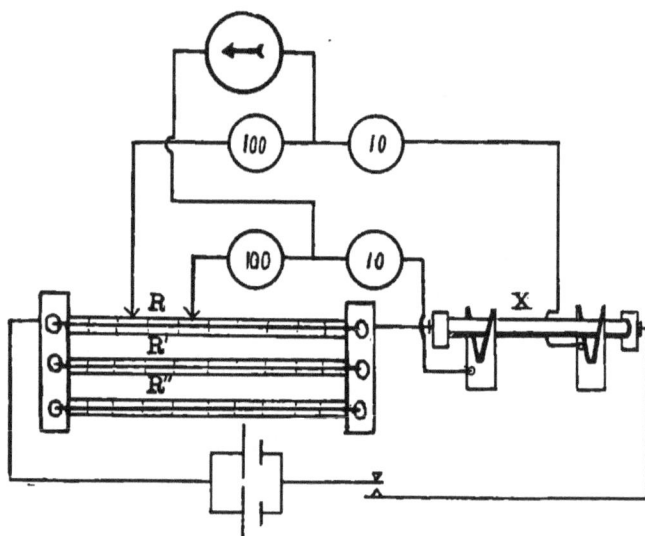

FIG. 13.

We then have three standard wires and make use of either one or the other, according to how low the resistance is that must be determined. The arrangement is best shown by the diagram.

It is evident that the measurement is only limited by the sensitiveness of the galvanometer and the strength of current that may be employed.

A D'Arsonval galvanometer or a Thomson high resistance galvanometer should be used. The Thomson low resistance galvanometer is too strongly effected by thermal currents, although it is far more sensitive.

Using a copper wire with a resistance of about .oocoo2 ohm
per mm., two "Samson" cells in parallel, and a D'Arsonval
galvanometer with a "sensitiveness" of only about 50 micro-volts,
it is possible (if a telescope and scale be used to observe the
galvanometer deflections) to make measurements to about
.oooooı ohm.

Now it is easy to obtain galvanometers with a sensitiveness
of 1 micro-volt, and twice the current strength may well be employed. The measurement could then be made to .ooooooo1
ohm. That is, a millionth of an ohm may be measured with an
accuracy of one per cent.

We may place the limit of what seems at present the lowest .

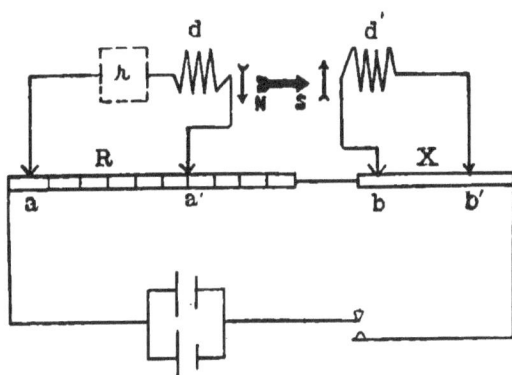

FIG. 14.

possible resistance measurement at about .ooocoooo1 ohm, or
the one-billionth of an ohm.

Differential Galvanometer.—A very similar method to the one
just described is that where a differential galvanometer is employed. In this case, it is possible that even considerably lower
resistances may be determined than with Thomson's double
bridge ; but a special form of galvanometer is required.

If the coils *d, d'*, Fig. 14, have an equal and opposite effect on
the galvanometer needle, then when the P. D. between *a, a'*, equals
that between *b, b'*, there will be no deflection, and the resistance
of R will equal X.

R is a standard wire or bar with adjustable contacts *a, a'*. The
resistance per scale division is accurately determined by the
step-down method given for the double bridge.

When it is desired to have the ratio of R to x as 10 : 1, a resistance, r, is added to the coil, such that $r + d = 10\,d'$, then there must be ten times the P. D. between a, a', that there is between b, b', before the current through the coil d will be the same as that through d'.

Hence, when R is adjusted to no deflection, R : x :: 10 : 1.

At the commencement of the measurement, whether the additional resistance is used or not, the coils d, d' should be tested for differentiality. This is accomplished by joining the wires a and b to some point of the circuit, and the wires a' and b' to some other point of the circuit.

If there be a deflection on closing the key, a small auxiliary resistance is added to one of the coils until there is no deflection.

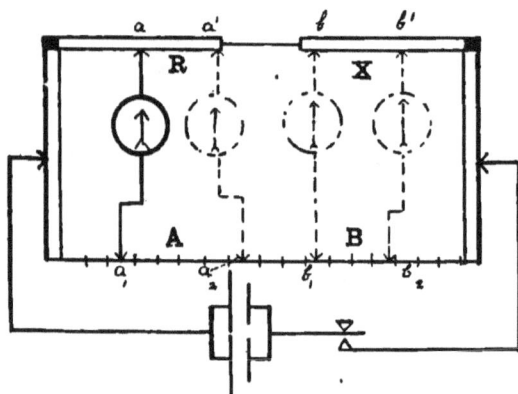

FIG. 15.

Projection of Potentials.—The very smallest resistances could easily be determined by this method, were it not for the many practical difficulties.

Let the standard resistance R and the unknown resistance x be joined in parallel with the bridge wire A, B, Fig. 15. One terminal of the galvanometer is joined to some point on the standard bar, and the other terminal a_1 is adjusted on the bridge wire A, B, to no deflection. Similar adjustments are made at the points a', b, b'; then A : B :: R : X. A and B are the lengths of the bridge wire included between the points $a_1\,a_2$ and $b_1\,b_2$.

The principle of the method is as follows : Since the point a, Fig. 16, is at the same potential as the point a_1, which is shown by no deflection of the galvanometer, and the point a' at the same dotential as the point a_2, there must be the same P. D. between

a, a' that there is between a_1, a_2, and likewise the same P. D. between b, b' that there is between b_1, b_2.

If these potential differences are represented by E, E', E_1, E_2, then R : X :: E : E', but E $=$ E_1 and E' $=$ E_2, hence R : X :: E_1 : E_2, but E_1 : E_2 :: A : B, $\therefore$ A : B :: R : X.

The method is an excellent one for moderately low resistances, provided that R be nearly of the same value as X; for if the ratio be very unequal, the length of the wire A, B, corresponding to the smaller resistance will not be great enough to obtain an accurate reading.

For very low resistances, the ratio of the leads and contacts to the resistances to be determined may become so great that but

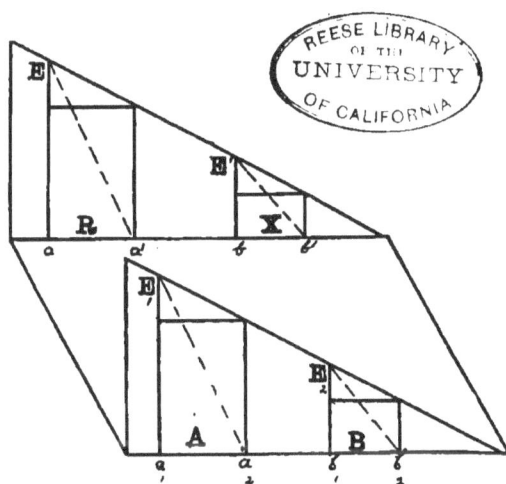

FIG. 16.

a small portion of the wire will remain available for the desired readings.

This difficulty may be partly overcome by adjusting the position of the battery terminals until the points, a_1 b_2, fall near the ends of the ratio wire.

Fall of Potential.—This method is exceedingly convenient for all ordinary measurements of low resistance.

A high resistance galvanometer is shunted across a known resistance, R, Fig. 17, and the deflection, d, is observed. The deflection d' across the unknown resistance, x, is then read, then $d : d'$ :: R : X. This assumes that the current is constant during the experiment. The resistance of the galvanometer should be great compared to the resistances R and x.   The readings should be taken quickly to avoid polarization and thermal effects.

If R be a standard wire or bar placed over a divided scale and the resistance per division be known, then R may be adjusted until $d = d'$ and we have R = x.

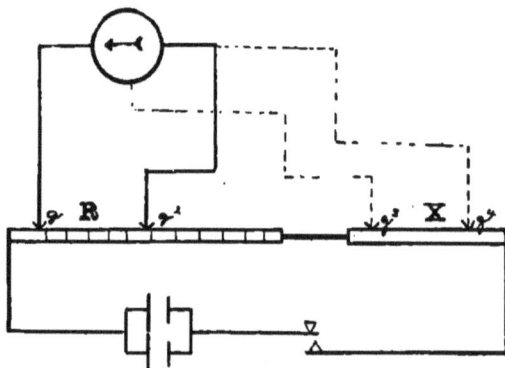

FIG. 17.

A voltmeter may be employed in place of the galvanometer where the resistances are not very low.

If the galvanometer has a sensitiveness of one micro-volt, and the resistance in the circuit is 0.5 ohm, then with a P. D. of two volts a deflection of one scale division would correspond to 0.25 micro-ohm.

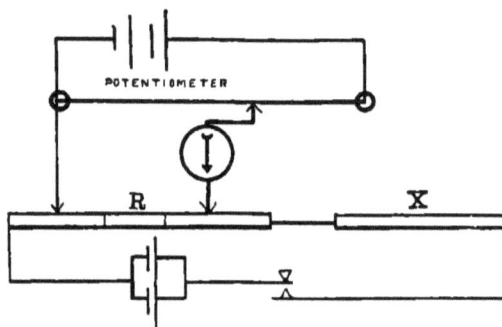

FIG. 18.

Potentiometer Method.—This method is simply a modification of the one just described.

The arrangement of the experiment is indicated by the diagram, Fig. 18.   The potentiometer reading is taken across R and

then across x. The readings are directly proportional to the resistances. The galvanometer should give no deflection when the
adjustment is correct.

Carey Foster's Method---Where the ordinary Wheatstone wire
bridge is employed for resistance work, this method may be
found convenient.

r_1 r_2 are approximately equal resistances, say one ohm coils.
x is the resistance to be determined, and b a copper contact
piece, Fig. 19. A balance is obtained in the position shown in the
diagram. x and b are then reversed and the galvanometer slider

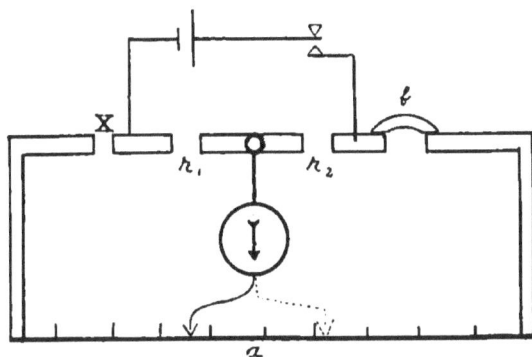

FIG. 19.

again adjusted to no deflection. The length of bridge wire a,
included between these two positions, is then equal in resistance
to the difference of resistance between x and b. The resistance
of the bridge wire per division may be determined by using,
say, 0.1 ohm in place of x.

The method is approximate and not applicable to very low
resistances.

It assumes that the contact resistances are the same in both
positions and that b is negligibly small compared to x.

CHAPTER IV.

THE WHEATSTONE BRIDGE.

All ordinary cases of medium resistance may be accurately and conveniently determined by the Wheatstone Bridge.

The demonstration of the principle of this method is similar to that given for the projection of equi-potentials.

Let R be a known resistance, x the resistance to be measured, and A and B resistances whose ratios are known, Fig. 20 ; then, if on closing the circuit there is no deflection of the galvanometer, A : B :: R : X.

Since the galvanometer shows no flow of current between g and g', the potential at these points must be the same, and since the other ends of R and A are joined the potential at this point must be the same. Consequently, there is the same P. D. across R that there is across A, and similarly the same P. D. across x that there is across B.

But from Ohm's law R : X :: P. D. across R : P. D. across X, also A : B :: P. D. across A : P. D. across B, and from the above, R : X :: P. D. across A : P. D. across B, or R : X :: A : B.

The Wheatstone Bridge may be termed an electrical balance in which the resistances compared are directly proportional to the balance arms.

The positions of the battery and galvanometer are interchangeable, but the greatest sensitiveness is obtained when that which has the higher resistance is placed at the junction of the two higher resistances.

The sensitiveness is also greater the more nearly equal are the resistances A, B, R, X.

The Wheatstone Bridge is used in a great variety of forms, but there are two general classes : That in which the ratio of A to B is varied while R remains the same, and that in which the ratio of A to B is given a constant value while R is varied until there is equilibrium.

The following is a list of these different forms of bridges :

VARIABLE RATIO..	Wire Bridge....	Straight (Metre.) Circular (Kohlrausch, Siemens. Parallel (Poggendorff.) Direct Reading (Kirchhoff.)
	Slide Coil Bridge.	Duplex (Varley.) Quadruplex (Muirhead.) Five Arc * (Cushman.)
CONSTANT RATIO..	P. O Bridge......	Elliot Pattern. Western El. Pattern.
	Decade Bridge...	Straight Pattern. Dial Pattern.

In the ordinary wire bridge, known as the metre bridge, or
B. A. bridge, the ratios A, B, consist of a G. S. wire, of which the

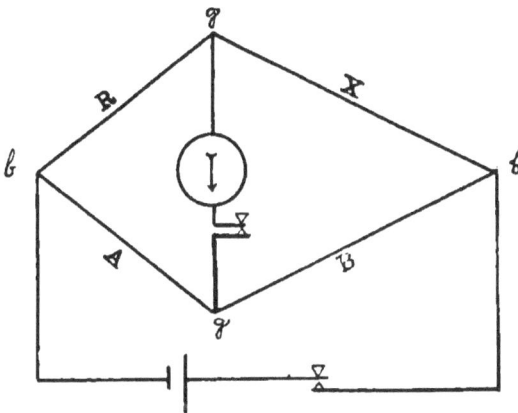

FIG. 20.

lengths may be accurately measured by means of a metre scale,
over which the galvanometer slider moves.

It is evident that the nearer the point g' is to the centre of
A, B, the less will be the effect on the measurement of any given
error in reading the scale.

The ends of the bridge are assumed to have no appreciable
resistance compared to the other resistances in the circuit ; but
if the resistance to be measured is very low, this is not true.
This resistance may be determined and corrected for, but since
the contact resistances cannot be eliminated, it is better to employ one of the special methods for very low resistance work.

This bridge is not suitable for very high resistance measurement, for the wire A, B, being of comparatively low resistance,
acts as a shunt either to the battery or galvanometer. The

ratio wire is often stretched around a divided circle ; this renders the birdge much more compact.

In the Siemens' bridge, a circular wire is used in combination with a set of standard resistances and a galvanometer.

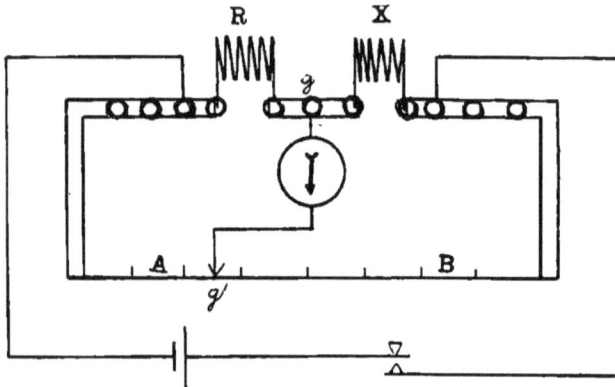

FIG. 21.

The Kohlrausch bridge has the wire wound on a drum ; by this means a considerable length of wire may be employed and very accurate readings obtained.

Another method of increasing the length of the wire is to have it stretched parallel as in the Poggendorff bridge.

A form of bridge, where instead of finding the ratio of A to B,

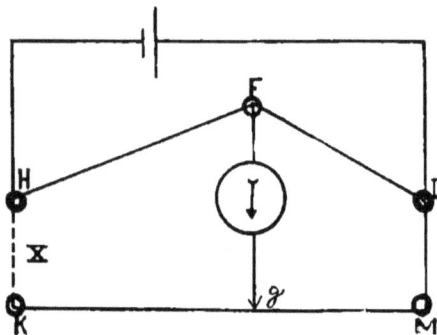

FIG. 22.

the resistance is directly read off, is known as·the direct reading bridge or Kirchhoff bridge.

Referring to the diagram, Fig. 22, it is necessary that the resistances should be in the following proportion :

$$M D : M K :: F D : F H.$$

Suppose these be given some definite value such as .3 : 3 ::
.2 : 2.   Then when x = 0, g will be at M; but if g is at K, then
3.3 : X :: .2 : 2, and x = 33.   Consequently the wire M K should
be divided into 33 parts.

Suppose a bridge reading up to 10 ohms is required, then

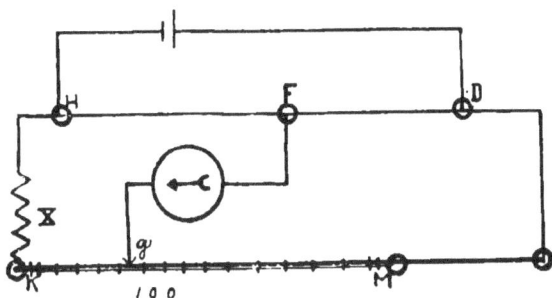

FIG. 23.

M K : M D :: F H : F D, and 10 : M K + M D :: F H : F D.

A laboratory form of this bridge is shown in Fig. 23.

In the commercial form of instrument known ss " Cardew's
lighting conductor bridge," a circular bridge wire is employed.

Slide Coil Bridges.—When high resistances are to be compared,
the ratio resistances A and B should also be high

In order to accomplish this, a series of coils may be used in

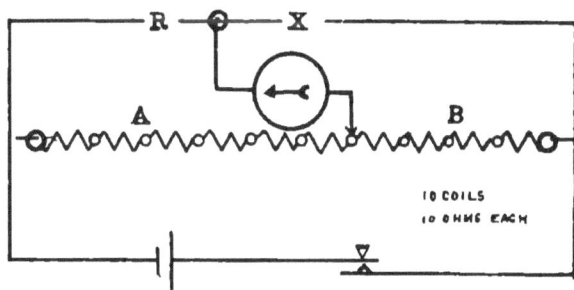

FIG. 24.

place of the bridge wire.   Suppose we have ten 10 ohm coils in
series, it would be equivalent to a bridge wire of 100 ohms resistance that could only be read to tenths, Fig. 24.

If, instead of this, eleven 10 ohm coils be employed, and a
second series of ten 2 ohm coils be arranged so that they may

be paralleled across any two of the first series, Fig. 25, the entire resistance will still be 100 ohms. The galvanometer slider may then make contact at any junction of the second series, and by this means readings obtained to $\frac{1}{100}$th. This arrangement constitutes practically an electrical vernier. The principle may be extended

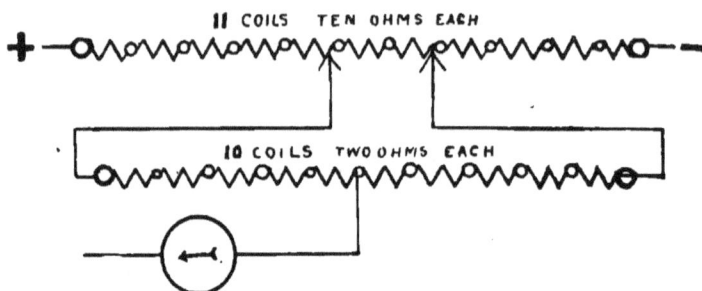

FIG. 25.

indefinitely in either direction. The following conditions should be observed: The entire resistance of the last series of coils must be equal to the resistance of two coils in the preceeding series. If there be 10 coils in the last series, the others must contain 11 coils, or if 100, then 101, etc.

The resistance of the coils in the last series should not be so low that they cannot be accurately adjusted.

The principle may be shown by referring to Fig. 26. Since the p. d. is proportional to the resistance, the p. d. from A to B is the same as that across 10 ohms. Consequently the p. d. across

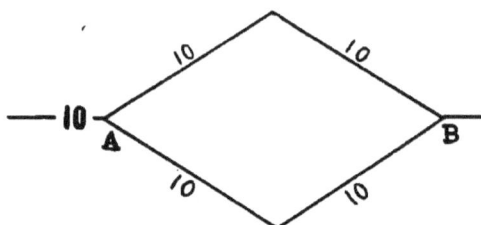

FIG. 26.

20 ohms in either of the parallel branches is the same as that across 10 ohms in the direct circuit. Therefore the p. d. across 2 ohms in one of the parallels is $\frac{1}{10}$ the p. d. across 10 ohms in the direct circuit.

Probably the best known form of slide coil bridge is the

Thomson-Varley instrument.   This employs 101 coils of 1,000 ohms, and 100 coils of 20 ohms, Fig. 27.

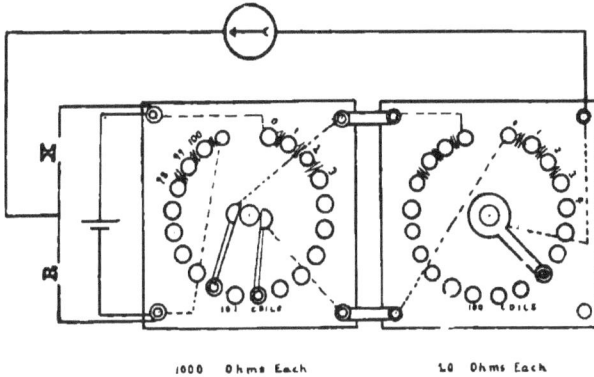

1000   Ohms Each   20   Ohms Each
FIG. 27.

The entire resistance is 100,000 ohms, and it gives readings to $\frac{1}{100} \times \frac{1}{100}$, or .0001.   The coils are arranged in two dials.

The Muirhead pattern has four series of coils, 3 of 11 each and 1 of 10 coils; the coils are arranged in dials.   The reading is: $\frac{1}{10} \times \frac{1}{10} \times \frac{1}{10} \times \frac{1}{10}$, or $\frac{1}{10000}$.   If a 10,000 ohm bridge is employed, the following would be the values of the resistances in the different series :

First.   1,000 ohms (11 coils).
Second. 200   "   "
Third.   40   "   "
Fourth. 8   "   (10 coils).

The Cushman five arc instrument, Fig. 28, possesses many advantages.   It employs :

11 coils of 10,000 ohms each.
11 "   "   2,000   "   "
11 "   "   400   "   "
11 "   "   80   "   "
10 "   "   16   "   "

The entire resistance is 100,000 ohms, while readings may be obtained to $\frac{1}{10} \times \frac{1}{10} \times \frac{1}{10} \times \frac{1}{10} \times \frac{1}{10}$, or $\frac{1}{100000}$.

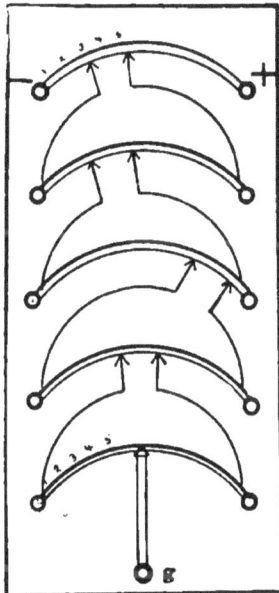

FIG. 28.

By this means the range of the bridge is increased 10 times

over that of the Varley Bridge, and but 54 coils are used instead of 201. With this instrument, resistances whose ratio is as great as 1000 : 1 may be compared with an accuracy of about one per cent. When the resistances are equal, of course, they may be compared to .002 per cent.; but errors due to temperature co-efficients, contacts, adjustment of the coils, etc., would probably be much greater than this.

Each series of coils is provided with binding posts, so that any of the higher series may be left out when a lower resistance bridge is desired.

For quick work, the settings of the lower series may be omitted.

The arrangement of the coils in arcs gives great ease and rapidity of adjustment.

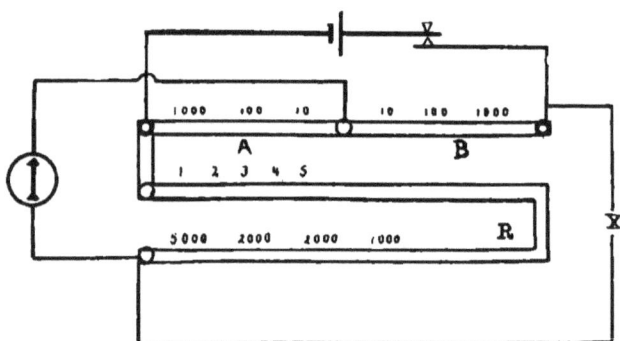

FIG. 29.

The slide coil bridges described above are generally known as " potentiometers " on account of their employment in the comparison of potential differences.

For most resistance work, the constant ratio form of Wheatstone bridge is to be recommended.

The two patterns of the P. O. bridge are shown in diagrams 29 and 30.

A and B may be given several different values, such as 1000 : 10 etc. R can be adjusted from one ohm to 10,000 ohms. Resistances are inserted by removing pegs.

The ratio of A to B is determined by the resistance to be measured. If x is very low, A should be as great as possible and B as small as possible, thus : 1000 : 10 : : R : x.

For higher resistances A and B should not be made so unequal that a change of one unit in the adjustment of R will not be shown by the galvanometer.

If the galvanometer does not show a change of several units in R, A must be made equal to B and as nearly equal to X as possible. Thus, suppose x is about 1,000 ohms, then we may have 1,000 : 1,000 : : R : X.

If x be very great, say about 1,000,000 ohms, then we must have 10 : 1,000 : : R : X.

Thus the range of these bridges is from .01 ohm to 1,000,000 ohms, though, of course, it may be increased by the use of additional resistance coils in A and B or R.

For accurate work in the measurement of low resistance, such as the determination of specific resistance, special manipulation is required.

To the binding posts $t\,t'$ Fig. 30, clips are added and the ends of the bridge are thus brought near together. The circuit is then

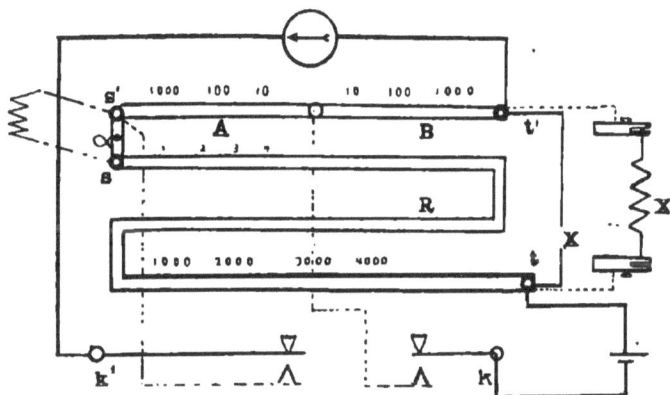

FIG. 30.

closed by inserting a wide, thick piece of copper between the clips. The $\propto$ peg is then removed and a piece of fine copper wire is joined to the posts s s'. The ratio of A to B is made 1,000 : 10. The length of the auxiliary wire is then adjusted until there is no deflection on closing the circuit.

The resistance of this wire compensates for the resistance of the peg row, contacts, clips etc. The copper plate between the clips is then removed and the resistance to be measured is joined to the clips.

With a sensitive reflecting galvanometer, resistances far below .01 ohm may be determined by interpolation.

The operation is as follows: Suppose, when R = 4 ohms, a deflection of 250 scale divisions to the left is obtained, and when

R = 5 ohms, a deflection of 150 divisions to the right is obtained. Then the exact value of R is 4$\frac{500}{800}$ ohms = 4.625 ohms, and 1000 : 10 : : 4.265 : x, or x = .0427 ohm.

At the beginning of the measurement, the galvanometer key k' should be closed ; if there is a deflection it is due to thermal currents. The battery key k is then closed, leaving the key k' open ; if there is a deflection, it is due to induction. These effects should be corrected for.

Generally, it is better to connect the battery to the posts $t\ k$, as indicated in Fig. 30. If the battery had a low internal resistance and were joined to the posts $t'\ k'$, in the case where the compensating resistance is being adjusted or where x is a very

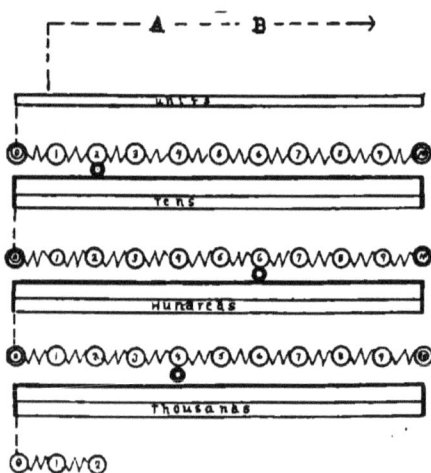

FIG. 31.

low resistance, it would practically be short circuited. Moreover, the heavy current thus obtained would heat the resistances and tend to destroy the balance.

The Decade pattern of Wheatstone Bridge involves the same principles of construction as those just described. But the arrangement of the coils is somewhat different.

This arrangement is shown for a portion of R in diagrams 31 and 32. There are 10 one ohm coils in series, 10 ten ohm coils, etc.; these coils may be placed in parallel rows or arranged circularly as in the dial pattern.

Each terminal marked zero is connected to the next preceding contact bar or circular contact block.

The number of coils thrown in circuit in any row or circle depends on the position of the peg in that particular row or circle.

In the dial form, a rotating contact piece may be used in place of a peg. The reading of R in the diagram given would be 462.

The advantage of this form of bridge is that the value of R may be read off directly, while in the P. O. bridge the resistances must be added up. There are also fewer pegs to manipulate, and where sliding contact pieces are employed the adjustment

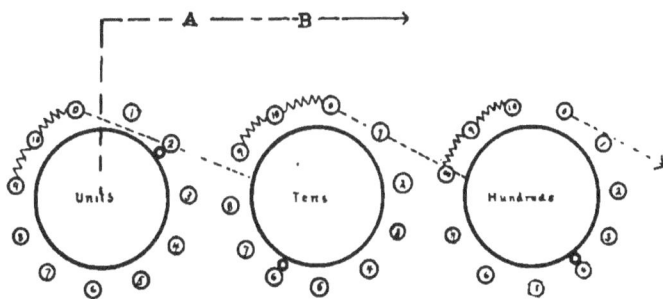

FIG. 32.

can be made with much more rapidity. The number of coils required in this style of instrument is, however, considerably greater than in the P. O bridge.

In a form of the Decade set known as the Anthony Bridge the resistance coils may be connected in parallel as well as in series. The lowest coils in R being one tenth ohm, when multiplied give $\frac{1}{100}$ ohm, and since the ratio of A to B may be made 10,000 : 1, the range of this bridge is theoretically from 1×10^{-4} ohm to 100 meg-ohms.

CHAPTER V.

Specific Resistance is the ratio of the resistance of any material to the resistance of any given material of the same dimensions under the same conditions taken as a standard.

The standard adopted is pure copper, and tables are made out giving the resistances of various lengths of copper wire of different diameters at several temperatures.

The best table is probably that adopted by the American Institute of Electrical Engineers. The data from which this table has been computed are as follows :

Matthiessen's standard one metre-gramme hard drawn copper
= 0.1469 B. A. ohms @ 0° C.

"   "   "   " soft drawn copper = 0.1437 B. A. ohms @ 0° C.

"   "   "   " soft drawn copper = 0.1417 international ohms @ 0° C.

Sp. Gr. of copper = 8.89.

Temp. coef. for 20° C = 1.0797. [.21 per cent. per degree F.]

One B. A. ohm = 0.9866 international ohms.

One Legal ohm = 0.9972 "   "

The table is made out in international ohms. One column gives the ohms per foot @ 20° C. of wire of various diameters. Thus, No. 20 wire, diameter = 0.03196 inch, resistance = 0.01014 ohm @ 20° C.

Since the resistance varies directly as the length and inversely as the square of the diameter. A single constant like the above might be used for the calculation of specific resistance.

The determination of specific resistance may be made as follows : A convenient length of the wire is taken and its resistance accurately measured on the P. O. bridge according to the directions given. The resistance thus found is divided by the length and the resistance per foot obtained. The diameter of the wire is then determined with micrometer calipers or wire gauge.

The resistance per foot is divided by the resistance per foot of copper wire of the same diameter, @ 20° C., the value being taken from the table.

For copper, the reciprocal of the above is taken and the result expressed in per cent. of conductivity. If the bridge employed is in B. A. ohms or legal ohms, the resistance should be reduced to international ohms.

If the temperature is much different from 20° C., a correction should be made.

In general, the resistance of alloys is much higher than that of any of their constituents, while the temperature coefficient is lower and may even be negative.

The Conductivity Balance is practically a Wheatstone Bridge, in which R consists of a standard copper wire of given length

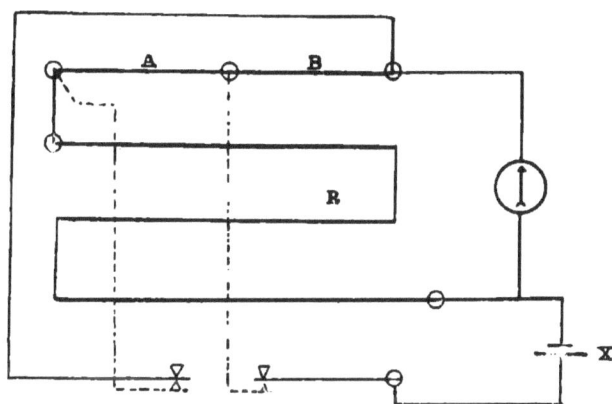

FIG. 33.

and weight. The length of the sample of wire to be determined is then varied until a balance is obtained, and from its length and weight the specific resistance is calculated.

It is scarcely less trouble than the ordinary method, and since the resistance of the standard may change, it is doubtful if results obtained can always be relied upon.

Galvanometer Resistance.—The resistance of a galvanometer may be measured by the P. O. bridge in the usual manner, but if it be a Thomson galvanometer, the coils should be turned until the needles are in the axis of the coils; the galvanometer, whose resistance is being measured, should be placed at such a distance from the other galvanometer, or in such a position with reference to it, that there will be no deflection produced by the inductive action of its coils.

When only one galvanometer is available, Thomson's method can be employed. The galvanometer is put in place of the unknown resistance in the P. O. bridge, Fig. 33, the ends of the bridge being joined by a wire furnished with a key.

The battery is joined between A and B, and R and the galvanometer. The ratio of A to B is determined by the galvanometer resistance.

The battery key is closed, and the galvanometer deflection is reduced to some convenient amount by lowering the magnet if it be a Thomson galvanometer, or adding an auxiliary resistance to the battery circuit. R is then adjusted until closing the key joining the ends of the bridge produces no change in the steady deflection.

Then, A : B :: R : galvanometer resistance.

One-half Deflection Method.—Another method is to join up the galvanometer in series with a rheostat and low resistance battery.

The resistance is adjusted until a convenient deflection is obtained. Call this resistance r. The resistance is then increased until the deflection is reduced to one-half. Call this latter resistance R. Then if the deflections are proportional to the current, the resistance in the second case must be twice that in the first case, since the current has been diminished to one-half. Hence :

$$2 (g + r) = g + R, \text{ or } g = R - 2 r.$$

If the battery resistance is appreciable compared to the galvanometer resistance, a correction should be made.

A modification of this method would be to shunt the battery until the desired deflection is obtained, and then add a resistance R, such that the deflection is one-half. Then, $g = R$.

CHAPTER VI.

Comparison of Standards.
Calibration of Bridge Wire and Rheostat.

In the comparison of standard ohms, an accuracy of from .01 per cent. to .001 per cent. is desirable. This means that the determination must be made to within at least .0001 ohm. Some special method must, therefore, be employed. The best is Carey Foster's Method.

In the diagram, Fig. 34, A, B, is a large German silver wire whose resistance per metre has been accurately measured, and from this the resistance per mm. calculated before placing the wire in position on the bridge. R, R', should be approximately one ohm each. They are known as the ratio coils. s, s', are the standard ohms to be compared. They are placed in water baths, and contact is made with the bridge by means of mercury cups.

The galvanometer slider is adjusted on the wire A, B, until there is no deflection.

The positions of s, s', are then reversed and a balance again obtained. The length of wire, r, included between the two positions of the galvanometer slider is equal to the difference in resistance between s, and s'. The slider is moved toward the greater resistance. Example : Resistance of A, B, per mm. = .00005 ohm, r = 3 mm., slider moved toward s'. Then s' is greater than s by .00015 ohm or .015 per cent.

The standards should be placed in the water baths a considerable time before commencing the measurement, and the temperature during the determination carefully noted. Several observations should be taken, and after some time another set obtained. If these last differ materially from the first, it is probably due to the standards not having arrived at a constant temperature.

The temperature coefficient of the wire composing standard ohms may be taken at about .02 per cent. to .04 per cent. per degree C., so that a difference of ½° C. would have accounted for the difference of resistance in the above example.

At the beginning of the measurement the galvanometer key

41

should be closed, the battery key remaining open. If there be a deflection caused by thermal currents, it is probably due to the ends of the bridge being at different temperatures.

It is, therefore, well to cover the bridge ends with cotton or some other non-conducting material, if the apparatus is not used in a room of constant temperature.

It should also be noted if there is any effect due to induction. This is observed by closing the battery key, leaving the galvan-ometer key open.

Where great accuracy is required, the utmost care must be exercised throughout the determination.

A special form of bridge, such as the Jenkin's bridge, is usu-ally employed. The instrument is extremely compact. A short standard wire is used, and copper blocks with mercury cups and commutator so arranged that the standard ohms to be compared

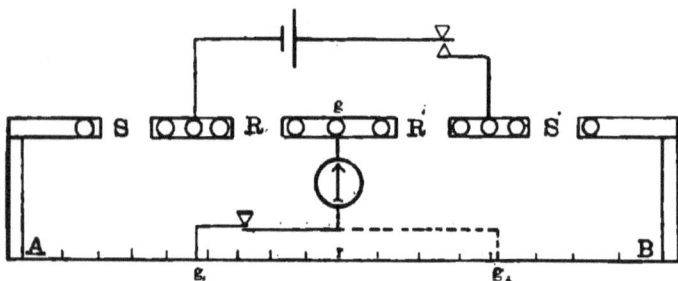

FIG. 34.

may be placed very near together and their positions reversed by means of the commutator. The ratio resistances are wound on the same bobbin and thus identity of temperature is in-sured.

Another method by which resistances may be compared with considerable accuracy is by "substitution in the bridge." In Fig. 35, A, B, is a German silver wire, preferably of about three or four ohms resistance. R is an auxiliary rheostat. I, an interpolation resistance that may be made .001, .01 or .1 ohm, and s, s', the resistances to be compared. At the commencement of the de-termination, s and I are short-circuited, the one ohm peg is re-moved in the rheostat, R, and a balance against the standard resistance, s', is obtained by moving the galvanometer slider, g', until there is no deflection.

An interpolation resistance of, say .001 ohm is then added and the galvanometer deflection observed. This resistance is then short-circuited and another observation made to see if the balance remains unchanged. For interpolation, it is convenient to

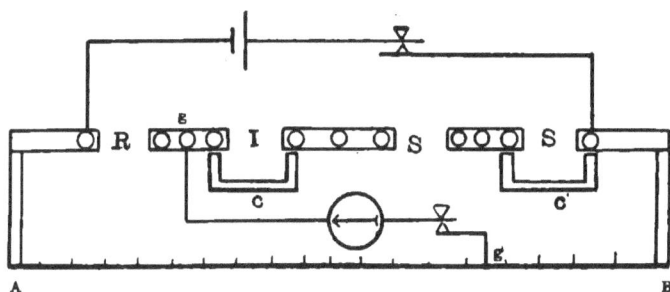

FIG. 35.

have an interpolation box, and add the resistances by removing pegs.

If the balance is still perfect, s' is shunted by the short-circuit piece c', and the resistance, s, to be compared is shown in circuit. The galvanometer deflection is then read and the difference in resistance calculated.

Example : Interpolation resistance = .001 ohm ; deflection = 160 (to the right) ; deflection when s is substituted for s' = 40 (to the left) ; then s is less than s' by $\frac{40}{160}$ of .001 ohm or .00025 ohm.

The same precautions with regard to thermal currents, etc., should be taken as in the first method.

For the comparison of standard ohms the Carey Foster method should be employed ; but where the resistances in a rheostat are to be checked against a standard resistance, the method just described is particularly applicable.

Calibration of a Bridge Wire.—The most direct and convenient manner of checking a bridge wire is to determine the lengths of wire that correspond to the ratio of known resistances.

Suppose R, R', Fig. 36, to be two rheostats each adjustable from 1,000 to 10,000 ohms. If it be desired to step off the bridge wire A, B, into 10 parts of equal resistance, R is first

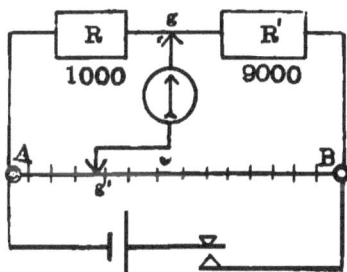

FIG. 36.

made 1,000 ohms and R' 9,000 ohms, and g' adjusted to no deflection. This point gives the first tenth. R is then made 2,000
and R' 8,000, and the second point on the bridge wire determined, and so on for the other points.

Of course, lower resistances for R, R', might be used if the
leads were negligibly small compared to them.

The resistances, R, R', are supposed to be as correct as the
percentage of accuracy required in checking the bridge wire.

It is important that the battery leads be connected directly to
the ends of the bridge wire A, B. In that case the resistance of
the leads from the rheostats is added to the large resistances
R, R', and tends but slightly to disturb their ratio.

In place of the two rheostats it is exceedingly convenient to
employ a slide coil bridge (or potentiometer.) In Carey
Foster's method, an auxiliary bridge wire A', B', Fig. 37, is made

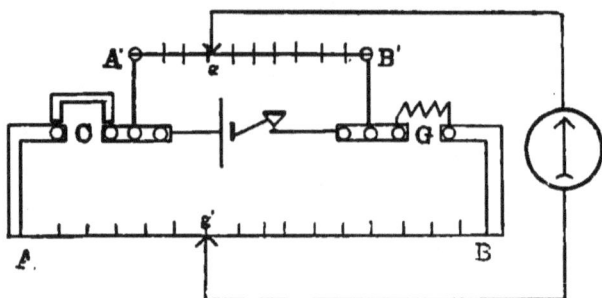

FIG. 37.

use of. G is the "gauge" or a resistance equal to the resistance
of certain length of A, B. according to the desired interval of
calibration. C is a copper connecting piece.

The operation is as follows : The "gauge" being in the position shown in the diagram, g' is placed very near the end of the
bridge wire at B. and the other slider, g, of the galvanometer, is
adjusted on the auxiliary wire A', B', to no deflection.

The positions of G and C are then reversed and the slider g'
adjusted to no deflection. G and C are replaced in their former
positions, the slider g adjusted to no deflection, and so on.

By this means the bridge wire A, B, is stepped off in portions
of equal resistance.

The resistance of the "gauge" determines, of course, the
amount of displacement of g' at each reversal. With the double
bridge, the wire may be calibrated by simply measuring any
convenient resistance on different portions of the wire.

When the differential galvanometer is made use of, the bridge wire may be checked by determining some suitable resistance at different points along the wire.

Calibration of a Rheostat.—A rheostat may be calibrated by the method of " Substitution in the Bridge," previously described.

The one ohm coil in the rheostat is compared to a standard ohm.

Then the one ohm + the standard is balanced against two ohms in the auxiliary rheostat. An interpolation resistance is added and the deflection noted. This is then short-circuited and the two ohm coil in the rheostat to be calibrated substituted for the one coil plus the standard, the deflection noted and from this the difference in resistance calculated.

The one ohm coil + the two ohm coil is next balanced against three ohms in the auxiliary rheostat, interpolation made, and the three ohm coil is then substituted for the two ohm coil + the one ohm coil, and the difference in resistance determined. By this means all of higher resistances in the rheostat may be compared with the sum of the next lower ones.

When the resistance in the circuit is increased, the deflection of the galvanometer for any given change of resistance grow less, so that it is necessary to interpolate after each change of resistance. It is better to have several resistances for interpolation, such as 0.1, 0.01 and 0.001 ohm. and when the deflections become small to use one of the higher resistances.

Since this is a deflection method, thermal and inductive effects should be carefully corrected for.

CHAPTER VII.

HIGH RESISTANCE.

{ Slide Coil Bridge.
{ Fall of Potential.*
{ Deflection Method.
{ Loss of Charge.

It is difficult to define any exact limit for the term " High Resistance." In a general way, any resistance above 100,000 ohms may be called a high resistance, though usually in insulation measurements the unit taken is a million ohms or a meg-ohm (Ω). Again, in certain cases of insulation such as that of a telegraph line, the insulation may be considerably less than a meg-ohm. The term " insulation " is applied to the resistance of dielectrics and materials that are not good conductors. Thus any "insulation " is usually a " high resistance," but any " high resistance " is not necessarily that of a dielectric or an " insulation."

Just how great a resistance can be measured with the present methods it is hard to say, but theoretically by the deflection method, using 200 volts and a Thomson galvanometer whose figure of merit is 100,000 Ω, resistances up to 20,000,000 Ω might be determined. The imperfect insulation of the apparatus, however, is likely to produce considerable error before reaching resistances nearly so great as the above figure.

The accuracy required in high resistance measurement is much less than for the lower resistance measurements, but in many cases of insulation the resistance of the dielectrics vary greatly according to circumstances, and it is therefore necessary to know the conditions of the experiment with great exactness.

Slide Coil Bridge Method.—Very great resistances could be measured on the Wheatstone Bridge if the other three arms could be made of such resistance that they would be somewhat near the magnitude of that to be determined. This require-

---

* " Fall of Potential " should be substituted in place of ' Potentiometer " in the general classification.

ment is fulfilled if the standard resistance be at least
100,000 ohms, and a slide coil
bridge of say 100,000 ohms be
used for the adjustable ratio.

The connections are shown
in Fig. 38. A rather high
voltage should be used.

By this method resistances
of several meg-ohms may be
determined with great accuracy.

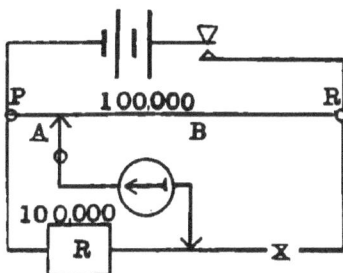

FIG. 38.

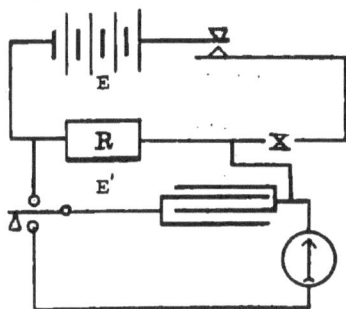

FIG. 39.

Fall of Potential.—Another
method is to join the known
resistance in series with the
battery and the resistance to
be measured. The E. M. F. of
the testing battery E having
previously been determined
with a voltmeter or by other
suitable means, the P. D. across
R (E') is measured. This is
best accomplished by charging a condenser across R, and
discharging it through a high resistance galvanometer. The value
of the deflection so obtained may be found by afterwards charging the condenser with a standard cell and noting the deflection.

The resistance of x is found by the proportion :

$$\text{E} : \text{E}' : : (\text{R} + \text{X}) : \text{X}.$$

The P. D. across R can also be measured with an electrometer
or a potentiometer and standard cell may be used ; in the latter
case the conditions of the circuit are rather uncertain, and it is
doubtful if the results obtained can be relied upon.

Deflection Method.—The standard method for nearly all insulation determinations is that of " direct deflection." Very great
resistances can be measured by it, the conditions of experiment
are completely under control and may be varied according to
circumstances.

A high resistance Thomson galvanometer is joined up in series with a known resistance and testing battery. The deflection
is then read, the galvanometer being shunted, and from this the
" constant " is calculated.

By the " constant " of the galvanometer is meant the resistance that must be introduced into the circuit to reduce the
deflection to one scale division with any given battery.

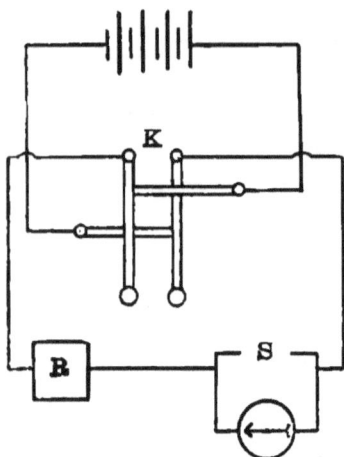

K

B

S

FIG. 40.

It will be seen that it depends upon the "Figure of merit" of the galvanometer and the E. M. F. used ; but since the same E. M. F. is employed throughout the measurements, it is not brought into the calculation. After the "constant" is obtained, the resistance to be measured is substituted in place of the known resistance and the deflection again read. This deflection, multiplied by the shunt, if one be used, and divided into the "constant," gives the resistance desired.

The connections are indicated by Fig. 40. Here a Kempe's reversing key is shown.

Example : R = 100,000 ohms (0.1 Ω), deflection with R in circuit = 250 divisions, when $\frac{1}{1000}$ shunt is used ; then "constant" = 0.1 Ω × 250 × 1,000 = 25,000 Ω ; deflection with x in circuit = 50 divisions, shunt = $\frac{1}{10}$. Then resistance of x = 25,000 Ω ÷ 50 × 10 = 50 Ω.

If the insulations to be measured are not very great, the Thomson galvanometer may be replaced by some form of D'Arsonval galvanometer.

Loss of Charge Method.—For very high insulations that cannot be conveniently measured by "direct deflection," this method may be employed. It is especially useful in testing joints of insulated wires. The connections for this determination are shown in Fig. 42.

A condenser is charged and the discharge deflection, v, noted. The condenser is again charged using the same E. M. F. and insulated for T seconds with the resistance to be measured between the poles ; it is then discharged and the deflection, v, noted.

Then if F = capacity of condenser in micro-farads, R = resistance in meg-ohms between poles of condenser.

$$R = \frac{T}{2.303\ F\ (log\ V - log\ v.)}$$

The resistance of the condenser should first be determined before placing the resistance to be measured between the poles. The final result is, of course, the combination of the two resistances when placed in multiple.

CHAPTER VIII

INSULATION. {
- *Insulating Material—*" Specific Insulation."
- *Insulated Wires.* {
  - Short Lengths.
  - Cable. { Single Core. / Multiple Core.
  - Joint Testing. { Loss of charge* / Accumulation. / Electrometer.
- *Aerial Wires.....*
}

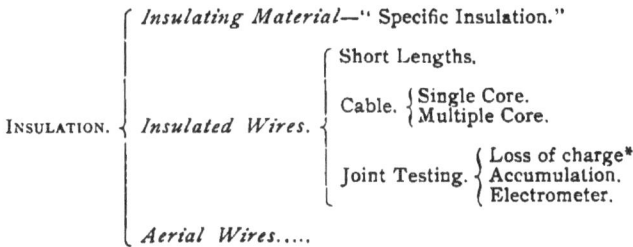

The requirements in the measurement of insulation are so varied that it seems best to make out a classification of the different cases and then treat each separately.

The resistance of dielectrics differs so greatly according to circumstances that a determination is of little value unless all of the conditions are given.

The E. M. F. used is of great importance ; it should usually be from 100 to 2co volts.

It is true that if the insulation is perfect it is independent of the E. M. F., but in a faulty insulation a high E. M F. may discover faults that a low one will not, hence a determination made with a low E. M. F. may be valueless.

The battery should be very constant, for if there is any capacity in the circuit, slight variations of E. M. F. may produce considerable variations in the galvanometer deflections.

Secondary batteries are the best. The silver chloride testing cells require careful handling to keep in good order and may get to have a very high internal resistance.

Insulation resistance decreases with an increase of temperature, and there is also a time lag. Hence, it should be kept at the same temperature for some time. The standard temperature is 75° F.

The resistance of dielectrics appears to increase by the continued action of the current. This action is known as "electrification." It seems to be due to a sort of dielectric polarization.

The deflection should, therefore, be read after some stated time—usually after one minute " electrification."

49

It is much more marked at low than at high temperatures. It depends on the kind of material, being quicker in some kinds of gutta-percha than in others, and is smallest in the best quality.

In the case of gutta-percha the rate of fall between the first and second minute would average about 2 per cent. to 5 per cent. In india rubber it may be as much as 50 per cent. be tween the first and fifth minute.

If the insulation is sound, the "electrification" should be regular. An unsteady "electrification" is usually a sign of defective insulation. It may be caused, however, by a bad condition or insulation of the battery, imperfect insulation of the ends of leads or cable, or it may be due to currents induced in cables when they are coiled.

There also seems to be a difference of effect in cable testing, whether the + or — pole of the battery is put to the cable. When the battery is not reversed, the — pole should be joined to the cable. The + pole seems to have the effect of sealing up a fault. When the current is reversed, a good insulation should give equal deflections.

Specific Insulation.—This should be determined by the deflection method, substituting the insulating material in place of R in Fig. 40. The insulation of the apparatus and leads should first be carefully tested. It is best to use a rather large surface of the insulating material. The contact may be made in the following manner: On a well insulated support or table is placed the wire leading from the galvanometer; next comes a piece of tinfoil the size of the area to be measured; then a piece of wet felt or wet blotting paper of the same size, and upon this the insulite. The contact above is made in the same way with the addition of a cover upon which is placed a heavy weight. By this means good contact is secured over the surface to be measured.

The deflection should be taken after one minute "electrification." This gives the "absolute" insulation. The "specific insulation" is the insulation of unit volume. It is obtained from the absolute insulation by multiplying by the area of the contacts and dividing by the thickness of the insulating material.
Thus:
$$\text{Sp. Ins.} = \text{Ab. Ins.} \times \frac{\text{area}}{\text{thickness}}$$

The E. M. F. used should be stated and the temperature at time of experiment.

Short Lengths of insulated Wire—When short lengths of cable, such as ¼ mile to one or two miles are to be tested during manufacture to obtain the insulation per mile, the following method should be employed.

The connections are shown in the diagram, Fig. 41.

The cable is coiled and placed in a tank of water. One end is carefully insulated while the other end is joined to the circuit. Contact with the inner surface of the insulation is thus obtained by means of the conducting wire, while contact with the outer surface is secured through the water into which attached to a metal plate dips the wire from the other end of the circuit. Fig. 41 is a general diagram for cable testing and does not show this special arrangement.

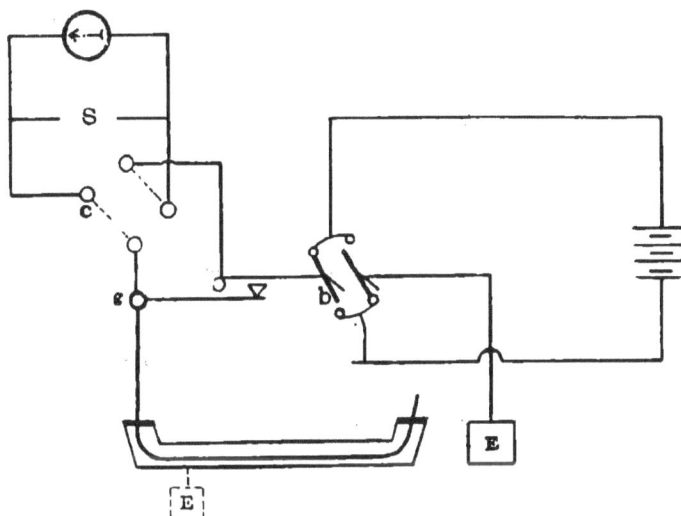

FIG. 41.

The galvanometer is provided with a commutator, c, and a short circuit key, g, the battery is joined to a key of the Rymer-Jones pattern. The cable is shown in position for testing. The manipulation is as follows : The battery key being closed, the short-circuit key, g, is opened and the deflections read at the end of one minute and two minutes ; g is then closed, the current reversed, the galvanometer commutated and the deflections again read.

The idea in commutating the galvanometer is to obtain the deflections on the same side of the scale. This is important when the deflections with first the — and then + poles of the battery joined to the cable are compared. Of course, the " constant " should also have been obtained from deflections on the same side of the scale.

The short-circuit key must always be used in cable testing, for cables act like condensers, having a capacity of about ⅓ mf. per mile, and are charged and discharged every time the battery key is opened or closed. This charge and discharge would, of course, take place through the galvanometer were it not shori-circuited.

In the report, the insulation after one minutes' "electrification" with the — pole joined to the cable should be given. The percentage of "electrification" between the first and second minute should also be reported.

It is better to keep the cable in the water for at least 24 hours before making the determination. The standard temperature for the water is 75° F.; at any rate, the temperature should be specified.

The e. m. f. used should be stated.

The "absolute" insulation divided by the length expressed in miles will give the insulation per mile.

A good cable should have an insulation of from about 400 Ω to 1,000 Ω per mile.

Cable Testing.—This is the case where the cable is in position, either sub-marine or underground. The connections are the same as Fig. 41. The terminal, E, is "grounded," and one end of the cable is insulated. The resistance after one minutes' "electrification," the — pole being joined to the cable, is reported and also the "electrification" between the 1st and 15th minutes.

In making careful tests on long cables, a rather complicated set of observations is required, and a report based upon these is made out.

In cases of this kind it is advisable to consult "Kempe's Handbook of Electrical Testing."

When submerged cables are tested, the "earth current" may render the readings unsteady. There is no method of eliminating these effects in single cored cables, but if the cable is multiple cored, a second core may be joined to the battery in place of "grounding" it and the total insulation divided by two.

Joint Testing.—Here the length of cable tested is very short and, consequently, the resistance is enormously great so that some more delicate method than "direct deflection" must be employed. One of the most convenient methods is that by "loss of charge." The connections are shown in Fig. 42.

The condenser is first charged and the discharge deflection

taken. It is then charged and insulated with the joint to be tested submerged in water in well insulated vulcanite trough be-tween the poles. The deflec-tion, after a certain time, is again taken, and difference between this and the first de-flections shows the loss of charge for the given time.

The rate of loss of charge through a perfect joint is first obtained for a fixed time, and afterwards the loss of charge for the same time through the joints tested is compared to this taken as a standard.

Another method is to charge the condenser for a certain time through a perfect core and note the deflection, and

FIG. 42.

then charge it through the joints to be tested.

An electrometer may also be used. The quadrants are charged through the joint and the deflection noted. This is then compared with a perfect core.

Aerial Wires.—In this case, the resistances to be measured are usually comparatively low, so that if the deflection method be employed less elaborate apparatus can be used, or the deter-mination may be made by the P. O. bridge.

One end of the wire is joined to one of the bridge terminals, the other terminal being "grounded." The other end of the wire should be insulated. The measurement can then be made in the usual manner. When the deflection method is used, the constant should be obtained in ohms instead of meg-ohms.

The standard insulation of the English Postal Telegraph De-partment is 200,000 ohms per mile. If the resistance is below this, the line is considered faulty.

The insulation per mile is approximately equal to the total insulation multiplied by the length in miles. It may be ob-tained accurately by the formula :

$$i = R_1 \frac{R_\circ}{\eta}$$

Where i = insulation per mile, R_1 = total resistance of line with one end insulated, $R_\circ$ = total resistance of line with further end grounded, η = the conductivity resistance of line per mile.

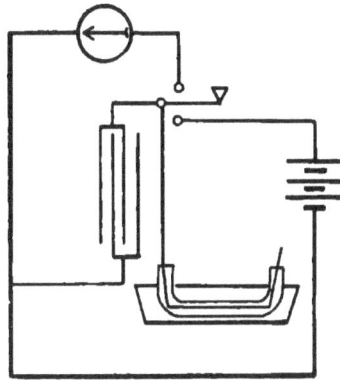

It is sometimes required to find the insulation resistance of two sections of one wire when it can only be tested from one end.

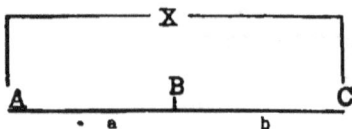

Suppose A, C, to be the wire which is required to be tested for insulation resistance from A in two sections, A B and B C. Let a be the insulation resistance of the section, A, B, and b that of B, C ; and suppose x to be the insulation resistance of the whole wire from A to C, then

FIG. 43.

$$x = \frac{a\,b}{a + b}$$

from which

$$b = \frac{a\,x}{a - x}$$

It is only necessary, therefore, in testing from A, to get the end c insulated and measure the insulation resistance, x. Then get the wire separated at B, the end of the section, A, B, insulated and measure the insulation resistance, a. From these two results b can be calculated.

CHAPTER IX.

RESISTANCE OF TELEGRAPH LINES, CABLES, ETC.

{
P. O. Bridge.
Loop Test. { a.
               { b.
Equilibrium.
Mance's Method.
Equal Deflection.
}

When the conductivity resistance of a wire is to be measured whose further end is not at hand, one end should be joined to one of the terminals of a P. O. bridge while the other bridge terminal is put to earth—the other end of the wire is also put to earth, and the measurement is then made in the usual way.

Whenever possible, however, it is better to measure without using a " ground," by looping two wires together at their further ends, the nearer ends being joined to the bridge terminals ; this gives the joint conductivity resistance of the two. Errors due to earth currents or a defective earth, etc., are thus avoided. The conductivity resistance of each wire separately cannot be obtained, however, by this means.

If there be three wires at hand, however, then the conductivity resistance of each wire may be obtained by making three measurements in the following manner :

Let the three wires be numbered respectively 1, 2 and 3. First loop wires 1 and 2 at their further ends, and let their resistance be R_1 ; next loop wires 1 and 3, and let their resistance be R_2 ; finally, loop 2 and 3, and let their resistance be R_3. Indicating the resistances of 1, 2 and 3 by r_1, r_2 r_3, we have $r_1 + r_2 = R_1$, $r_1 + r_3 = R_2$, $r_2 + r_3 = R_3$. From this, by adding the equations, we get

$$\frac{R_1 + R_2 + R_3}{2} = r_1 + r_2 + r_3 = R_4, \therefore r_1 + R_4 - R_3, r_2 = R_4 - R_2, \text{ and } r_3 = R_4 - R_1.$$

By a method very similar to the above, if there be only two wires at hand, the resistance of the " earths " at the ends of the lines may be measured and also the resistance of each wire.

Denote the resistances of the two wires by r_1 r_2, and the resistances of the " earths " by E.

First loop the wires, then $r_1 + r_2 = R_1$, next " ground " the first wire and measure the resistance, then $r_1 + E = R_2$; finally, ground the second wire, then $r_2 + E = R_3$. From these we get:

$$\frac{R_1 + R_2 + R_3}{2} = r_1 + r_2 + E = R_4,$$

then $E = R_4 - R_1$, $r_1 = R_4 = R_3$, and $r_2 = R_4 - R_2$.

Such a test, however, although it eliminates errors due to defective earths, does not eliminate errors due to earth currents.

When it is not possible to make use of the loop test and the conductivity, resistance of a line of telegraph must be obtained by " grounding " one end, the presence of earth currents and also currents due to the polarization of the earth plates, renders the Wheatstone bridge formula A : B :: R : x, when equilibrium is produced, incorrect. To obtain the true value of the wire resistance different methods and formulæ are necessary.

Equilibrium Method.--Fig. 44 shows the Wheatstone bridge arrangement, where x = resistance of wire to be determined, E = E. M. F. due to earth currents, etc., and r = resistance of the battery circuit.

R is first adjusted to no deflection ; call this resistance R_1. The current is then reversed and R readjusted to no deflection ; call this resistance R_2. Then by applying Kirchhoff's laws,

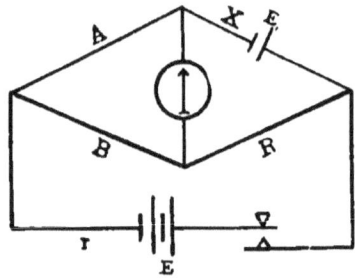

FIG. 44.

$$X = \frac{A}{B}\left[\frac{2\,(R_1 + k)\,(R_2 + k)}{(R_1 + k) + (R_2 + k)} - k \right]$$

where

$$k = \left[r\left(1 + \frac{B}{A} \right) + B \right]$$

Mance's Method.--Here the current is not reversed but the values of A and B are changed and R is again adjusted to no deflection. Call these values A_1, B_1, R_1, and A_2, B_2, R_2. In practice, $A_1 = B_1$, and $A_2 = B_2$;

where

$$X = \frac{R_1\,(2\,r + A_2) - R_2\,(2\,r + A_1)}{(R_2 + A_2) - (R_1 + A_1)}$$

Equal Deflection Method.--This is the same as Mance's method for the measurement of battery resistance except that a battery is included in the circuit joining the ends of the bridge. The connections are shown in Fig. 44. If there be an earth current the galvanometer is permanently deflected if the galvanometer key is kept closed. R is adjusted until on closing the battery key no change in deflection is produced. Then, A : B : : R : X.

In practice, it would be necessary to short-circuit the galvanometer at the moment when the battery key is closed or opened, otherwise a violent deflection of the needle would be produced by the static discharge from the cable.

CHAPTER X.

LOCALIZATION OF FAULTS.

The conditions met with in testing for the location of faults are so varied and complex that it is hardly possible to give any general method of procedure. The best way is to consider each of the several cases separately.

Faults may be of the following descriptions :

> 1. *Complete Fault in Insulation.*
> 2. *Partial Fault in Insulation (Earth Resistance).*
> 3. *Variable Fault in Insulation (Polarization or Cable Current).*
> 4. *Faults plus E. M. F. (Earth Current).*
> 5. *Fault in Conductor.*
> 6. *Faults of High Resistance.*

(1) The simplest kind of fault to localize is a complete fracture where the fault offers no resistance. Its position is easily determined by dividing the conductivity resistance to the fault by the conductivity resistance per mile of the line.

(2) When the fault has a resistance, the localization becomes more difficult.

The following are two theoretical methods that may be employed.

Blavier's Method.—Let A B, Fig. 45, be the line which has a fault f at c, A being the testing station. The end B is first insulated and the resistance from A to the fault measured ; call this l. The end B is then put to earth, and the resistance from A again measured; call this l_1. Then if the conductivity resistance of the line be L, and the resistance from A to c is denoted by a, it may be shown that

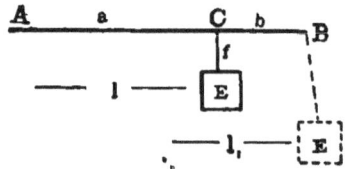

FIG. 45.

$$a = l_1 - \sqrt{(l - l_1)(L - l_1)}.$$

58

Dividing a by the resistance of the line per mile would, of course, give the position of the fault.

Overlap Method.—In this method two measurements are made, one from station A when B is insulated, and the other from B when A is insulated.

Calling the first resistance l, the second l_2, resistance of line ι, and the resistance from A to the fault a, then

$$a = \frac{L + l - l_2}{2}.$$

(3) The practical application of the above methods, however, presents considerable difficulty. This is owing to the variation of the resistance of the fault when the testing current is put to the cable, due to electrolytic action at the fault which may increase the resistance and also set up a polarization current in the opposite direction to the testing current. This is especially true in the case of sub-marine cables.

To make a proper measurement, then, it is necessary to so manipulate the testing apparatus and battery as to get rid of the polarization and resistance set up.

In Lumsden's method, the further end of the cable being insulated, the conductor is cleaned at the fault by applying a zinc current from 100 cells for 10 or 12 hours, the current being occasionally reversed for a few minutes. This followed by other special manipulation.

In Fahie's method, the cable current is tested by an auxiliary galvanometer. This current is then neutralized with a battery. If the cable current is negative, a positive current should be used in measuring the resistance.

(4) The principal difficulty in testing for faults is the presence of earth currents, especially in the case of long cables. These earth currents are seldom constant either in strength or direction for any length of time.

Mance's Method has for its object the elimination of the effects of an earth current in a cable when making a resistance test. The general principle of this method is the same as that previously described for the measurement of the resistance of a telegraph line. The bridge arms A and B are made equal, and then given two different values, corresponding adjustments being made for R.

In the *Deflection Method*, the Wheatstone bridge is not made use of.

A Thomson galvanometer with a reversing switch and shunt and a battery with reversing key are joined in series with the cable and earth. The testing current is reversed, the galvanometer being also reversed so that the deflections may be in the same direction. Since in one case the battery current is in the same direction as the earth current, and in the other case it is opposing it, the two deflections will differ. Call these deflections d_1 and d_2. A rheostat is then substituted for the cable, and the resistances adjusted until the same deflections are produced.

Call these resistances R_1 and R_2, and X the resistance to be measured ; then,

$$X = \frac{d_1 R_1 + d_2 R_2}{d_1 + d_2}.$$

This method, of course, requires careful manipulation.

When possible, the " Loop Test," described below, should be made use of.

(5) When the conductor is broken inside the insulating sheathing of a cable, a battery joined to the end of the cable will charge the latter up as far as the fault only. Consequently, if the discharge be measured and compared with the discharge from a condenser of known capacity charged by the same battery, the capacity of the cable up to the fault will be obtained. This capacity divided by the capacity per mile of the cable will give the distance of the fault.

(6) In the previous methods described for localizing faults, it was assumed that the insulation resistances of the portions of cable on either side of the fault were infinitely great, compared with the resistances of the conductor.

This assumption is practically true when the cable under test is short, and also if the resistance of the fault is small ; but in the case of long cables having faults of high resistance, the formulæ given above are no longer correct.

The latter case requires the use of very complicated formulæ.

Whenever possible, however, the loop test, with corrections, should be employed.

For a complete discussion of the subject of localization of faults, Kempe's " Handbook of Electrical Testing " should be consulted.

The Loop Test.—When a faulty cable is lying in the tanks at a factory, so that both ends of it are at hand, or when a submerged cable can be looped at the end farthest from the testing station, with either a second wire, if it contains more than one wire, or with a second cable which may be lying parallel with it, then the simplest and most accurate test for localizing the position of the fault is the loop test.

This test is independent (within certain limits) of the resistance of the fault.

There are two ways of making this test with the P. O. bridge.

Murray's Method.—(The connections are shown in Fig. 46.)

p is the point where the two wires or cables are looped together, f being the fault.

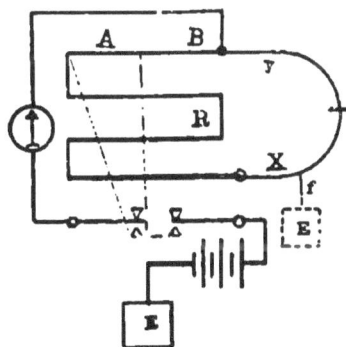

Let x be the resistance from one end of the bridge to the fault, y the resistance from the other end of the bridge to the fault. Then the arm B being plugged up, A and R are adjusted until equilibrium is produced. Then, $A \times y = R \times x$.

Let L be the total conductivity resistance of the whole

FIG. 46.

loop; then, $x + y = L$; therefore, $y = L - x$; substituting, $A (L - x) = R \times x$, from which $x = L \dfrac{A}{A + R}$.

L may be determined in the usual manner for measuring resistance.

A should be given a rather high value, say 1,000 ohms, in order that the range in adjustment of R may be increased.

The zinc current should be put to the cable (through the bridge).

The value of x divided by the conductivity resistance of the cable per mile or foot, as the case may be, gives the position of the fault.

Varley's Method.—(Fig. 47 shows the arrangement of this method.) A and B are fixed resistances, and R is adjusted until equilibrium is produced. Then, $B (R + x) = A y$, and $y = L - x$; therefore, $B (R + x) = A (L - x)$, from

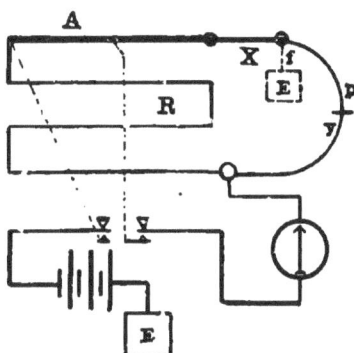

which $x = \dfrac{A L - B R}{A + B}$. If then $A = B$, $x = \dfrac{L - R}{2}$.

The conditions for making this test with accuracy are not quite so simple as in Murray's

FIG. 47.

method.   In this case they are almost precisely similar to what they are in the ordinary Wheatstone bridge measurement.

CHAPTER XI.

RESISTANCE OF BATTERIES AND ELECTROLYTES.

BATTERY RESISTANCE.
- *Fall of Potential.** { Condenser. / High Resistance. / Voltmeter.*
- *Added Resistance.* { Tangent Galvanometer. / ½ Deflection.
- *Mance's Method.*
- *Current and E. M. F.**

The resistance of a battery is not a perfectly fixed quantity. It may vary somewhat according to the strength of current that is flowing and the time that the current has been maintained. The resistance varies also with the temperature of the battery.

It is therefore desirable in an accurate determination to know the conditions of the circuit, especially the value of the external resistance.

The best method is probably that by *Fall of Potential.* This measurement may be carried out in several ways.

Condenser Method.—The cell or battery x is joined in series with a known resistance R and a key b. Across the terminals of the cell are connected a condenser, galvanometer and discharge key, shown in the diagram, Fig. 48.

The condenser is first charged by closing the key a, the key b remaining open. The condenser is then discharged through the galvanometer. The deflection thus obtained, d_1, is proportional to the E.M.F. of the cell.

The key b is then closed and the condenser again charged

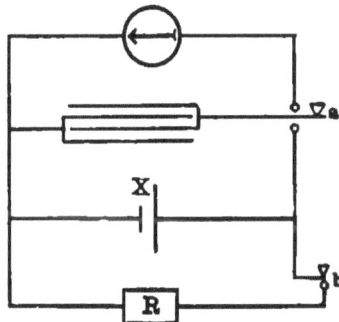

FIG. 48.

and discharged. This deflection, d_2, is proportional to the P.D. across R. Therefore $d_1 - d_2$ is proportional to the P.D. across x. Then:

$$d_1 - d_2 : d_2 :: x : R.$$

If R be varied until $d_2 = \dfrac{d_1}{2}$, then R = x.

R may be given different values, and the corresponding values of x determined, or x may be measured after the current has been flowing for different lengths of time. This method is especially applicable when the efficiency of any particular cell is to be investigated.

High Resistance Method.— A galvanometer and high resistance may be substituted in place of the condenser. The high resistance is joined in series with the galvanometer, and readings are taken across the terminals of the cell similar to those in the above method. The calculation is the same.

Voltmeter Method.—A low reading voltmeter, such as the Weston voltmeter, that can be read directly to $\frac{1}{10}$ volt and estimated to

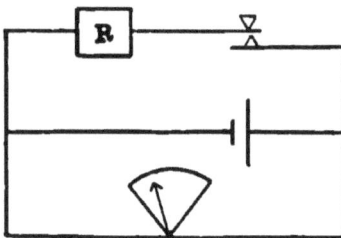

FIG. 49.

$\frac{1}{100}$ volt, may be used to take the deflections across the cell. Readings are made with the key open and with the key closed, in the same manner as in the condenser method, and the same calculation is used.

If R is adjusted until the deflection with the key closed is one-half of the deflection with the key open, then R is equal to the resistance of the cell.

This method is to be recommended for all ordinary determinations of cell resistance.

It is, of course, inapplicable in the case where the resistance of a battery is appreciably great compared to the resistance of the voltmeter.

Tangent Galvanometer Method.—If a low resistance tangent galvanometer be at hand, the determination can be made in the following manner : Take the deflection with the cell and galvanometer in series, call this deflection d_1, and the resistance of the circuit x. Then add a known resistance, R, preferably such that will about halve the deflection, call this d_2. Then :

$$x : x + R :: \tan d_2 : \tan d_1,$$

and   x — galvanometer resistance = the cell resistance, assuming the resistance of the leads to be negligible.

This method may sometimes be found convenient for a laboratory determination where the cell is fairly constant.

If a low resistance reflecting galvanometer be used and the resistance neglected, and such a resistance, R, added, that $d = \dfrac{d_1}{2}$, then R = the battery resistance.

If the battery has a fairly high resistance and the galvanometer resistance, G, is not neglible, then the following method may be used : The galvanometer, battery and a rheostat are joined in series. The resistance, R_1, is adjusted until some convenient deflection is obtained. Then the resistance is increased until the deflection is halved ; call this second resistance R_2. Then :

$$R = R_2 — (2 R_1 + G),$$

where R is the resistance of the battery.

Mance's Method.—In this method, the cell is joined to the terminals of a Wheatstone bridge, in place of the unknown resistance, the ends of the bridge being connected by a wire furnished with a key. The connections are the same as those shown in Fig. 44, except that no testing battery need be employed. The position of the cell is indicated by x.

The galvanometer key is closed and the steady deflection produced by the cell is reduced to some convenient amount either by lowering the magnet, if the Thomson galvanometer is used, or by the addition of an extra resistance to the galvanometer circuit. R is then adjusted until no change is produced in the deflection on closing the key joining the ends of the bridge. The manipulation in this method is sometimes rather difficult, and it is hardly to be recommended when fall of potential methods can be employed.

Current and E. M. F.—In some instances, where the internal resistance of a cell is very low, the P. D. across an external resistance subtracted from the E. M. F. of the cell gives such a small difference, that measurements cannot be made accurately by fall of potential methods.

In this case, it is best to measure the current with an ammeter and the E. M. F. with a voltmeter ; then the resistance can be calculated from the formulæ $R = \dfrac{E}{C}$.

RESISTANCE OF ELECTROLYTES. $\begin{cases} \textit{Contant Current.} \\ \textit{Alternating Current.*} \end{cases}$

When the resistance of a fluid, which is decomposed by the current, is to be measured, account must be taken of the opposing E. M. F. of polarization. The simplest method is that of substitution.

The fluid is placed in a U tube provided with platinum electrodes, one arm of the tube being calibrated.

The fluid thus contained is included in a simple circuit with a rheostat, a galvanometer and a galvanic cell.

The arrangement is shown in Fig. 50.

The position of the needle is then observed when so much of the column of fluid is included that the deflection is a convenient amount ; then one electrode is approached to the other by the length *l*, and such an amount R of rheostat resistance thrown into the circuit that the same deflection is produced. The resistance R is then equal to that of the fluid between the two positions of the movable electrode, assuming the polarization to be the same in both cases. It is well to use spirals of platinum wire or platinum gauze for the electrodes.

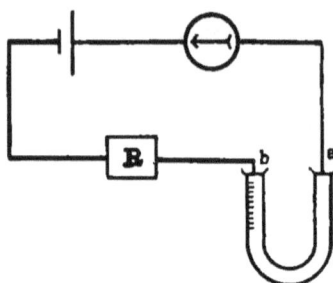

FIG. 50.

Since the conductivity of fluids varies greatly with their temperature, this should be observed, and be kept constant by placing the tube in a water-bath provided with a thermometer.

The influence of polarization may be avoided, and the resistance of an electrolyte measured directly, just as that of a metallic conductor, if a rapidly alternating current be employed.

The tube containing the fluid, x, is joined to the terminals of a Wheatstone bridge, the current being furnished by an induction coil. A telephone receiver is used in place of the galvanometer. The connections are shown in Fig. 51. The adjustment to equilibrium is obtained when the telephone gives the minimum sound, then A : B :: R : X. A number of observations should be made.

The electrodes in this case should consist of platinized platinum foil.

To obtain the "Specific Resistance" of a fluid, the containing vessel should first be filled with mercury, and

FIG. 51.

the resistance measured in the usual way. This gives the "mercury constant" of the vessel. This resistance divided into the resistance of the fluid gives its resistance compared to mercury, from which the resistance compared to copper can be

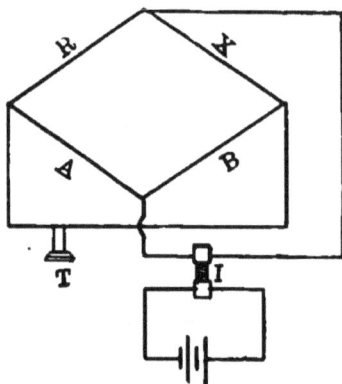

calculated. If the dimensions of the containing vessel, or rather the column of fluid measured, were accurately known, then the resistance of unit volume could be calculated.

CHAPTER XII.

INCANDESCENT LAMPS, " DYNAMO
RESISTANCE," ETC.
$\begin{cases} \textit{Fall of Potential.} \\ \textit{Current and E. M. F.*} \\ \textit{Ohmmeter.} \end{cases}$

It is often required to measure the resistances of incandescent lamps or arc lamps while running. The resistance of an incandescent lamp depends very largely upon the temperature of the filament, and consequently upon the strength of current flowing. It is therefore desirable that the original conditions of the circuit be interfered with as little as possible when the measurement is made.

The fall of potential across the lamp may be measured with

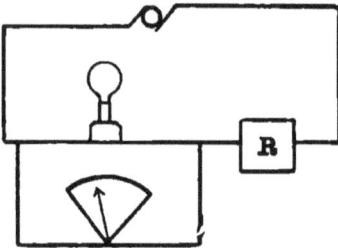

FIG. 52.

a voltmeter, and then the P. D. across a small resistance, such as an ohm, in series with the lamp (Fig. 52). The resistance is then calculated by a direct proportion.

In place of the voltmeter, a galvanometer and high resistance or a galvanometer and condenser can be employed.

The series resistance should be of such size wire that it will not be heated appreciably by the current, and the resistance should be small compared to that of the lamp, so that the current will not be materially reduced by it.

A better practical method, however, is to measure the current flowing through the lamp with an ammeter (an instrument having a negligibly small resistance), and the fall of potential across the lamp with a voltmeter. The resistance is then calculated from the formula $R = \dfrac{E}{C}$.

The resistance can also be measured directly by means of an ohmmeter. In principle, the ohmmeter consists of two coils at right angles to each other, with a small needle at the point of intersection of the axis (Fig 53). One of the coils of low resistance is in series with the resistance to be measured, and the other, which is of comparatively high resistance, is in shunt.

Under these circumstances the action of the needle is due to the ratio of the difference of potential at the terminals of the unknown resistance and the current strength in the series coil, or, $R = \dfrac{E}{C}$.

The coils are so proportioned, that when the current flows through the short thick wire, it moves the needle to the zero of the scale, while the long thin wire of the shunt coil produces a deflection directly proportional to the resistance.

If the coils are large and the needle short, the instrument will follow the tangent law.

In Evershed's ohmmeter, the current coils are wound outside and the shunt or pressure coil is globular in form, so as to fit inside. It is placed at an angle of 45°, so as to give a long scale. Inside the shunt coil a hard steel needle is suspended by a silk fibre. A second needle is hung outside the coils, so that the instrument is astatic.

The range of the instrument is increased by inserting resistance in series with the shunt coil. It is graduated by experiment.

An ohmmeter should always be tested to see if it is accurate. A piece of thick wire is measured in the ordinary way, and the resistance

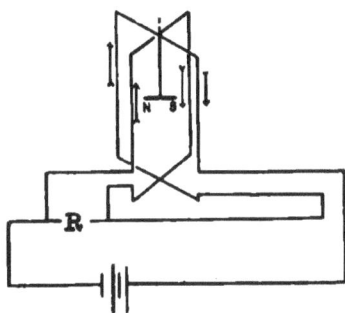

FIG. 33.

then determined with an ohmmeter. Care must be taken that the wire does not become heated. The same resistance should be measured with a large current and a small current.

The apparent resistance of a dynamo while running may be determined in the following manner : A voltmeter being connected across the terminals, the P. D. is measured on closed circuit. This gives the fall of potential across the external resistance, or the P. D. across the "line". The current should also be measured with an ammeter.

The circuit then being opened, the voltmeter indicates the total P. D. that the dynamo is capable of giving.

The first reading subtracted from this shows the P. D. across the dynamo when running on the above circuit. The resistance can then be obtained from the formula $R = \dfrac{E}{C}$. Of course, if the resistance of the external circuit were known, the resistance of the dynamo could be calculated without using the ammeter.

CHAPTER XIII.

Determination of the Ohm, Construction of Standards, Etc.

The measurement of resistance consists of the comparison of the resistances to be determined with some other resistance taken as a standard.

The derivation and determination of this standard is of interest, though, of course, its absolute determination is never required in practice.

That originally taken as a convenient unit was approximately the resistance of a mile of copper wire of a certain size. An exact value was assigned to it by Siemens, who defined it as the resistance of a column of mercury one metre in length and one square millimetre in section at the temperature of melting ice. This has been called the Siemens unit.

With the application of the c. g. s. system to electrical measurement and the adoption of the magnetic definitions, the unit of resistance was defined as the ratio of the centimetre to the second. The product of this quantity by 10^9, called the ohm, was designed for practical use, the c. g. s. unit being inconveniently small. It has nearly the same value as the Siemens unit.

The exact determination of this unit has been the work of many years. The original b. a. ohm of 1864 has been replaced by more accurate determinations of the c. g. s. unit. The Paris Conference in 1884 agreed upon the so called legal ohm, and defined the resistance as that of a column of mercury, 106 centimetres long, of one square millimetre section, at the temperature of melting ice.

At the British Association meeting of 1892, the results of a large number of independent measurements were compared, and what is now known as the international ohm or the true ohm was adopted.

Its resistance is defined as that of 14.4521 grammes of mercury in the form of a column of uniform cross section 106.3 centimetres in length at 0° C. (It is equivalent to a cross-section of one square millimetre).

The above value was also adopted by the Electrical Congress at Chicago in 1893.

Table I. shows the relative values of the different units.

An outline of the general method followed in the absolute determination is indicated below.

A coil of wire of known area and number of turns is placed so that the axis of the coil is in the magnetic meridian, and rotated with a known velocity. By definition the c. g. s. unit of e. m. f. is that obtained when one "line of force" is cut per second, and hence the e. m. f. developed by the coil can be calculated. Let F = the total strength of field ; then since each line of force is cut four times in one revolution of the coil, the e.m.f. = area of coil × number of turns × 4 × F × number of revolutions per second. The coil is joined in series with a tangent galvanometer (Fig. 54), and the current strength obtained by the formula

$$C = \frac{R}{2\,n\,\pi} \times H \times \tan B,$$

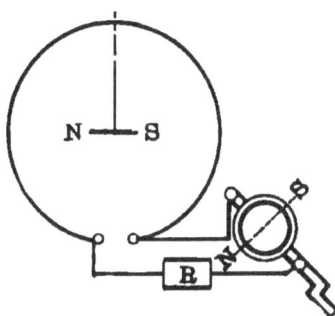

FIG. 54.

where B = deflection, R = radius of galvanometer coils, and n = number of turns.

The resistance of the circuit can then be calculated from the formula $R = \dfrac{E}{C}$.

TABLE I.

	Siemens Unit.	B. A. Ohm.	Legal Ohm.	International Ohm.
Siemens Unit	1.00	.9535	.9434	.9407
B. A. Ohm	1.0488	1.00	.9894	.9866
Legal Ohm.	1.c6	1.0107	1.00	.9972
International or True Ohm....	1.063	1.0136	1.0028	.1.00

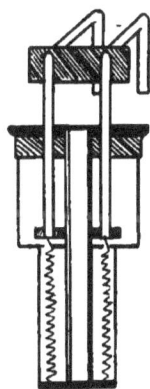

FIG. 55.

If a resistance be added to the circuit, the above operation may be repeated and the resistance of the circuit again obtained.

The difference between these two results would give the value of the resistance added.

Since the resistance is measured in c. g. s. units, it must be divided by 10^9 to reduce it to ohms.

The secondary standards consist of coils of wire of various alloys, whose resistance are known to be very nearly constant.

A form of standard resistance coil devised for the first British Association committee, and which was till recently the generally accepted form, is shown in Fig. 55.

It consists of a coil of wire on a metal bobbin with a tubular core, the ends being connected to a pair of thick copper rods, led through ebonite clamps, and bent downwards so as to be easily put into mercury cups.

The whole coil is then slipped into an outside case of thin sheet metal in the form of two cylinders. The lower cylinder contains the wire coil, and the upper is filled with paraffin wax. The case up to the shoulder is intended to be placed in a bath of water, the temperature of which is taken with a thermometer placed in the central tube after the coil has been so long in the bath that it has reached the temperature of the water.

The coils of a rheostat should be wound bifilar (Fig. 56) to neutralize the magnetic action of the current and prevent effects due to self-induction. This winding is most easily accomplished from two bobbins, the farther ends of the wire being soldered together. The coils should be dipped in melted paraffin to secure more perfect insulation and prevent any atmospheric action on the wire that might change the resistance in the course of time.

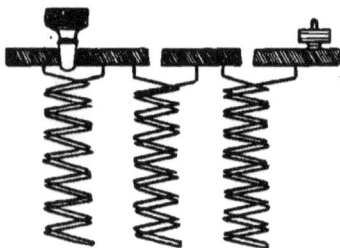

FIG. 56.

The following are the principal materials that have been employed commercially for constructing resistances.

German silver, an alloy of copper, nickel, and zinc. Specific resistance about 18, but varies greatly, according to the composition of the alloy. Temperature coefficient .04 per cent. per 1° C. Largely used in the construction of rheostat coils.

Platinum silver, composed of two parts by weight of platinum to one of silver. Specific resistance about 15. Resistance increases .031 per cent. per degree centigrade. Used for standards.

Platinoid is German silver with the addition of a small percentage of tungsten. Specific resistance about 17. Temperature coefficient .022 per degree centigrade.

Manganin, specific resistance about 20. Alloys containing manganese have been found to have very small temperature coefficients, and it is even possible to obtain them with negative coefficients. In the case of this, or any of the new alloys, a long series of observations are required to establish with any degree of certainty the permanency of the resistance.

CHAPTER XIV.

{ Electromotive Force and Potential Difference.
{ Measurement of E. M. F. of Batteries and Direct Currents.

The term "electromotive force" is not a scientifically accurate one, since in the Newtonian sense, force is only that which acts on matter.

This restricted use of the term "force," however, does not seem advisable when the present state of scientific theories with regard to the ether is considered. It would be better if force were defined as that which produces or tends to produce motion, or that which produces stress.

It may be well to consider briefly the possible nature of electrical action, in order to show more clearly the relation between electromotive force and potential difference.

Suppose some cause, an electromotive force, produces a stress in the ether. This stress, under certain conditions, produces an ether strain or displacement, and the change from stress to strain must, of course, be accompanied by motion, either vibrational or progressive.

This motion is known as an electric current. Its presence can only be recognized when the ethereal motion has been converted into the motion of the grosser particles of matter.

Since the same effect may be produced by different causes, though any given cause must always produce the same effect, potential difference may be due to a great variety of electromotive forces.

We may then define electromotive force as the cause which produces potential difference, and potential difference as that which produces or tends to produce an electric current.

From the above considerations it seems to the writer that the term "electromotive force" is a particularly good one ; it is only its mode of use that should be objected to.

It is evident that, strictly speaking, electromotive force can never be measured ; it is only potential difference that may be determined. It should also be remembered that in the statement of Ohm's law, $C = \dfrac{E}{R}$, that the E stands for difference of potential, and not electromotive force.

There has been a great want of uniformity in the employment of the term "electromotive force·" By some, it is regarded as that which causes difference of potential; others consider it as being produced by potential difference; and still others regard it as the entire electric moving cause produced by any source; while anything less than this is called potential difference. This last distinction between the two terms is that ordinarily used with regard to dynamos and batteries.

Whenever the term "electromotive force" (E. M. F.) is used with respect to measurements, it should be understood to indicate "total potential difference."

The abbreviation T. P. D. has been suggested in place of E. M. F. Its employment would save much confusion.

BATTERIES AND DIRECT CURRENTS.
- *High Resistance Method.**
  - Deflection and Resistance.
  - Equal Deflection.
  - Equal Resistance.
- *Wheatstone's Method.*
- *Lumsden's Method.*
- *Condenser Method.**
- *Potentiometer.**
  - Five Arc (Cushman).
  - Quadruplex (Muirhead).
  - Duplex (Varley).
- *Current and Resistance.*
- *Electrometer.*
- *Voltmeter.**

For the determination of potential difference (P. D.) in direct current and battery work, a great variety of methods may be selected from. The more important of these are indicated in the above classification. Of course, the most exact measurement is obtained by means of the potentiometer, but other methods are often better suited for special cases.

High Resistance Method.—If a source of potential difference, for example the battery E_1, is joined in series with a galvanometer and a resistance (Fig. 57), the E. M. F. is proportional to the deflection $d_1 \times R_1$, where R_1 equals the entire resistance of the circuit, for from Ohm's law $E = C R$.

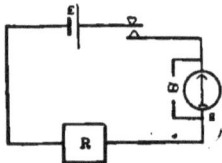

Let another battery E_2 be used in place of E_1, then the E. M. F. is proportional to $d_2 \times R_2$, where d_2 equals the deflection, and R_2 the entire resistance in circuit.

FIG. 57.

Hence, $E_1 : E_2 :: d_1 R_1 : d_2 R$.

The difficulty with this method is that it requires the resistance of the batteries and galvanometer to be known.

A modification of the above is to vary the resistance in circuit until the two deflections are equal.

Then, $E_1 : E_2 :: R_1 : R_2$.

If the resistance in circuit is equal in both cases, then

$$E_1 : E_2 :: d_1 : d_2 .$$

This is accomplished in practice by making the resistance R so great that the battery resistances, r_1 and r_2, are negligibly small compared to it. Then the E. M. F. is directly proportional to the deflection.

A sensitive reflecting galvanometer should be used, and the resistance R ought not to be less than 10,000 ohms, a resistance of 100,000 ohms being preferable.

Wheatstone's Method.—In this method, the battery E_1 is joined up in series with a galvanometer and rheostat ; a deflection d_1 is obtained. The resistance is now increased by R_1, so that the deflection is reduced to d_2. The battery E_2 is next used in place of E_1, and the resistance in circuit is adjusted until the deflection obtained equals d_1. The resistance is now increased by R_2, so that the deflection is reduced to d_2, as in the first instance ; then

$$E_1 : E_2 :: R_1 : R_2 .$$

That is, the E. M. F.'s are directly proportional to the added resistances.

Lumsden's Method.—This method may sometimes be found convenient in comparing the E. M. F. of battery cells.

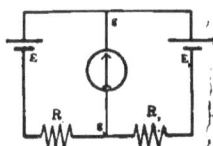

FIG. 58.

The two batteries E_1, E_2 are joined up, with their opposite poles connected together, and with resistances R_1, R_2, in circuit (Fig. 58); a galvanometer is connected between the points $g_1 g_2$. One of the resistances, say R_1, being fixed, the other, R_2, is adjusted until the galvanometer gives no deflection ; then $E_1 : E_2 :: R_1 : R_2$.

The method of making the connections when a P. O. bridge is used, is shown in Fig. 59.

If in place of making R_1 equal to 1,000 ohms, it be made some multiple of E_1, then when R_2 is adjusted so that no deflection is obtained, it must be equal to the same multiple of E_2. Thus, suppose $E_1 = 1.44$ volts, $R_1 = 1,440$ ohms, then if $R_2 = 1,079$ ohms, $E_2 = 1.079$ volts.

The above adjustment could be made by removing the a

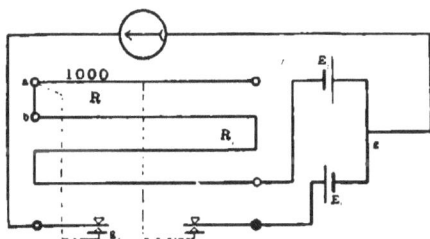

FIG. 59.

peg between posts *a* and *b*, and interpolating a resistance of 440 ohms, the galvanometer with an extra key in circuit being joined to the post *b*.

Condenser Method.—One of the most convenient as well as one of the most universally applicable methods is that in which the condenser is used. The resistance of the condenser is practically infinite as far as any other resistance in circuit is concerned, so that the E. M. F. of batteries, whose internal resistance is very great, can be accurately determined by this method. Again, there is not the slightest chance for any polarization effects to take place during the measurement.

The connections are shown in Fig. 60. When the discharge key K is depressed, points 1 and 2 are connected, the condenser F being charged by the battery E_1. When the points 1 and 3 are connected, the condenser is discharged through the galvanometer. The deflection d_1 is proportional to the capacity of the condenser F $\times$ E_1.

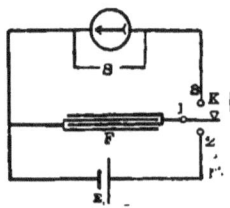

FIG. 60.

Another battery, E_2, is joined up in place of E_1, and the throw of the galvanometer d_2 again obtained, and since this deflection is proportional to F $\times$ E_2, we have:

$$E_1 : E_2 :: d_1 : d_2.$$

That is, the E. M. F.'s are directly proportional to the galvanometer deflections.

Potentiometer Method.—This method is the standard for the accurate comparison of E. M. F.'s, such as the checking of standard cells against each other, and also for the calibration of voltmeters.

It admits of the very greatest precision of adjustment, and is a zero method, that is, it does not depend on galvanometer deflections. Moreover, this method is practically a static one, for when the final balance is obtained, there is no flow of current from the branch circuit containing the cell to be tested. Thus the measurement is not effected by the internal resistances of the cells or batteries compared, however high, and is also uninfluenced by the addition of any resistance that may be placed in series with them to prevent polarization during the first adjustments.

The method depends on the law, that if a source of P. D. be joined to the ends of a resistance, the P. D. across any portion of the resistance is proportional to the resistance itself. Also, if the ends of a second circuit with a given E. M. F. be joined by a portion of the above resistance, such that the P. D. across it is equal to this E. M. F. and in the opposite direction, there will be no flow of current from the second circuit.

The connections are shown in the diagram, Fig. 61.

A constant battery B, whose E. M. F. is somewhat greater than that to be determined, is joined to the ends of a resistance R R',

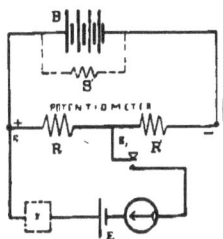

FIG. 61.

so arranged that the portion R included between the galvanometer terminals g_1 g_2 may be varied until the P. D. across R is just equal to the E. M. F. of the cell E, shown by no deflection of the galvanometer. This may be accomplished by employing a bridge wire and varying the position of the galvanometer slider, g_2, until there is equilibrium. For most work, however, it is better that this resistance should be high, at least 10,000 ohms, so in place of the bridge wire two rheostats may be used in the positions indicated by R and R'. The sum of these resistances R and R' must be kept equal to 10,000 ohms, that is, whenever R is increased, R' must be decreased by the same amount, and vice versa.

A far more convenient arrangement is to employ one of the slide coil bridges, known as "potentiometers," previously described. The positive pole of the battery is joined to the zero terminal of the potentiometer, and the negative to the opposite terminal. The positive pole of the cell to be tested is also joined to the zero terminal, and the negative to the slider or key, g_2, through the galvanometer.

The reading of the potentiometer, when adjusted to no deflection, shows the value of R, and the entire resistance of the potentiometer is equal to R + R'.

r is a resistance placed in series with the cell E, to prevent polarization during the trial adjustments, and may be short-circuited before the final adjustment.

s' is a shunt resistance joined to the terminals of the battery, in order to reduce the P. D. across R R' to any given amount, and need only be used when it is desired to make the potentiometer direct reading.

The determination is made in the following manner : A cell E of known E. M. F. is first used, and the value of R found when there is no deflection. A second cell, E_1, is then substituted the value of R, again found when there is equilibrium. Call this second value of R equal to R_1. Then,

$$E : E_1 :: R : R_1 .$$

The value of R and R_1 are given directly by the potentiometer readings.

In the comparison of standard cells, great care should be taken that they are of the same temperature and several readings should be made.

To make the potentiometer direct reading, a cell of known E. M. F., say 1,434 volts, is used, and R is made equal to it, in this case 1,434. s^1 is then adjusted until the galvanometer shows no deflection. Then if, when another cell is used, R is found equal to 1,078, the E. M. F. would be 1.078 volts.

Current and Resistance.—This method is of special use in checking standard cells and voltmeters.

In the diagram, B is a constant battery, R a known resistance, say 10 ohms, the wire being large enough not to be heated appreciably by the current. A is a current measuring device, such as a voltmeter, Thomson balance, or calibrated ammeter. s^1 is an adjustable resistance. E is the standard cell to be checked in series with a galvanometer, and *r* is a resistance to protect the cell from polarization. The key being closed,

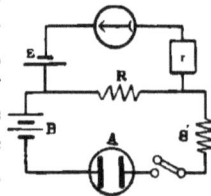

FIG. 62.

s^1 is adjusted until the galvanometer gives no deflection and the current C is measured. Then, E = P. D. across R = R C.

If a voltmeter is to be checked, it is joined across R, and the reading E taken with the key closed, the current C being measured at the same time.

If the voltmeter is correct, of course E should equal R C.

Electrometer.—The E. M. F. of cells can be quite accurately compared with the Thomson quadrant electrometer, if the needle be given a static charge and the cell terminals connected to alternate pairs of quadrants, Fig. 63. The cell is then connected and the deflection read ; call this *d.* Another cell, E^1, is substituted and the deflection d^1 measured. Then,

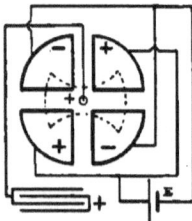

FIG. 63.

$$E : E^1 :: d : d^1.$$

A circular scale should be used. It is difficult to keep the needle at a constant potential unless a replenisher is employed and external electric charges are apt to effect the measurement.

The method is not to be recommended except for research work.

Voltmeter.—In a very great number of cases, the voltmeter is by far the most convenient means of measuring P. D.

A large class of voltmeters, practically all that are used for direct current and battery work, are really current instruments ; that is, the deflection is produced by the current flowing through the instrument. The scale, however, is calibrated to indicate the P. D. across the terminals of the voltmeter. The higher the resistance of the voltmeter, the more nearly will it give the true P. D. in any case where it is used in series, and the less will it tend to disturb the relation of the circuit when it is used in parallel. An example may, perhaps, make this plainer. Suppose a battery has a resistance of 100 ohms and an E. M. F. of 125 volts, and a voltmeter with 2,000 ohms resistance is used. Then the voltmeter, if correct, would indicate $\frac{2,000}{2,100} \times 125 = 119$ volts.

It is therefore well to know the conditions of the circuit and the resistance of the voltmeter in any particular measurement.

The portable and laboratory forms of the Weston voltmeter are well-known standard instruments and may be thoroughly relied upon to indicate correctly the P. D. across ther terminals. Fig. 64 gives an idea of their construction.

The principle is that of the D'Arsonval galvanometer. The magnet is placed horizontally, and the coil, in an oblique position between the pole pieces, turns in jeweled bearings. The current is lead in by means of watch springs. Two resistance coils, R, r, are placed in series with the movable coil. With R in series, the resistance in some of the instruments is about 15,000 ohms, and scale reads in volts

FIG. 64.

up to 150. With r in series, the resistance is about 500 ohms, and scale indicates $\frac{1}{30}$ volt up to 5 volts.

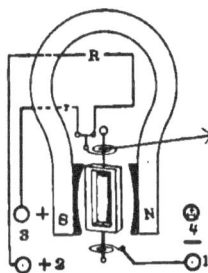

In order to change the reading, it is only necessary to use a different terminal.

The instrument is also provided with a commutator.

The following is a list of some of the more important forms of voltmeter, though, of course, nearly any device used for ammeter may be employed for a voltmeter, if it be given a sufficiently high resistance.

VOLTMETERS.

ELECTROSTATIC INSTRUMENTS.
- *Electrostatic Voltmeter* { Thomson's—gravity. / Weston's—spring.
- *Multicellular*—spring.
- *Low Reading Voltmeter*.

CURRENT INSTRUMENTS.
- *Weston's* .... { Direct Current Voltmeter—spring–magnet. / Alternating Current Voltmeter—dynamometers.
- *Ayrton & Perry's*...... { (*a*)—spring. / (*b*)—magnet.
- *Thomson's Graded Galvanometers*—magnet.
- *Magnetic Vane*—spring.
- *Eversheds*—gravity.
- *Cardew*—expansion.

CHAPTER XV

E. M. F. of Alternating Currents.—The E. M. F. of alternating currents may be conveniently measured by the following instruments :

ELECTROMETER. ⎧ *Electrostatic Voltmeter.** ⎰ Thomson's.
⎪ ⎱ Weston's.
⎨ *Multicellular.**
⎪ *Quadrant.*
⎩ *Low Reading.*

DYNAMOMETER. ⎰ *Weston's** (Alternating Current Voltmeter.)
⎱ *Siemen's.*

ATTRACTION OR ⎰ *Hartman and Braun's.*
ELECTRO-MAGNETIC ⎨ *Evershed's.*
VOLTMETERS. ⎱ *Magnetic Vane, etc*

CALORIC VOLTMETER (*Cardew's.*)

The form of electrometer known as Thomson's Electrostatic Voltmeter is very largely used in alternating current work.

The construction is shown by the diagram Fig. 65.

A light aluminium vane is pivoted on knife edges between two brass plates and is carefully insulated from them. The vane is provided with a pointer and to the lower end of the vane small counter weights may be added. The terminals of the source of P. D. are connected to the brass plates and the movable vane, the attraction is then proportional to the $\overline{\text{P. D.}}^2$. The scale is graduated in divisons proportional to potential differences. Three counter weights are provided and according to which is used the scale divisions are equal to 50 volts, 100 volts, or 200 volts respectively. The range of the instruments is thus very large,

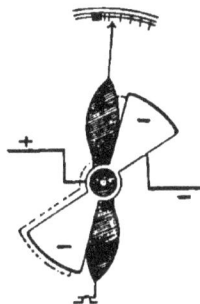

FIG. 65.

but it is liable to spark if more than 10,000 volts are used. It is
provided, however, with a safety fuse.

In the Multicellular Electrometer, Fig. 66, a
number of movable vanes are employed turning between corresponding fixed plates. The
force of attraction is balanced by the torsion of
the wire suspending the vanes. The scale is
graduated directly to volts.

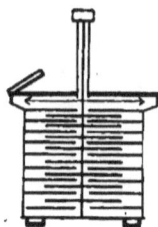

The instrument is made in four different
ranges, giving readings from 40 to 800 volts.
It is possible with this form of electrometer to
read as low as 15 volts.

FIG. 66.

The ordinary Thomson Quadrant Electrometer may be used
to measure an alternating E. M. F. if the vane and one pair of
quadrants be joined to a terminal from the source of P. D. and
the other terminal connected to the opposite pair of quadrants.
The connections are then reversed by means of a commutator
and the deflection observed. The E. M. F.'s are then proportional
to 2 √deflections.

This form of instrument is somewhat difficult to use and requires considerable care in the adjustments.

A special form of electrometer in which the moving vane is
rectangular in shape and suspended by a fine wire, is known as
the "Low Reading Electrometer." By means of a lamp and
scale, it is possible that as low an E. M. F. as $\frac{1}{10}$ volt can be measured with this instrument.

The employment of some form of electrometer, whenever
possible, is to be most strongly recommended.

Since it measures P. D. entirely by electrostatic pressure, its
use produces no change in the relations of
the circuit, and if the scale is graduated
correctly it must indicate the true potential difference.

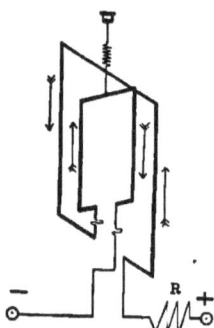

The *dynamometer* consists of a fixed and
a movable coil of wire, the latter being
normally at an angle to the plane of the
former, Fig. 67, and both coils being
traversed by the current whose E. M. F. is
to be measured.

FIG. 67.

Directive force may be given to the
movable coil either by the elasticity of a
spring or the torsion of a suspending wire.

The deflections of the movable coil are proportional to the
square of the current strength, and consequently to the square
of the E. M. F.

fastened to an axle provided with a counter weight and indicating hand, is placed with the coil to one side of the axis.   When

When a current passes through the coils, an attraction is exerted, and the movable coil tends to take a position parallel to the plane of the fixed coil.   '

Change of direction of the current in the entire instrument does not alter the direction of the deflection, and hence it is suitable for alternating work.

For small E. M. F.'s the dynamometer is not very sensitive, since the deflection is proportional to the square of the E. M. F.

For determination of E. M. F. the coils should be given a high resistance, or a high resistance should be placed in series with them.

The Weston alternating current voltmeter is a form of dynamometer in which the movable coil is mounted in bearings and the current lead in by watch springs fastened to the axle, the arrangement being similar to that described in the direct current voltmeter.   This instrument is quite sensitive and the readings reliable.   It is extremely useful for all ordinary measurements of alternating E. M. F.'s   It is possible, however, that the readings may be slightly effected by the action of strong magnetic fields.

In a form of the Siemen's dynamometer, the attraction between the coils is measured by the angle of torsion of an elastic spring, by turning the torsion circle of which the deflected coil is brought back to zero.   The E. M. F. is then proportional to the square root of the angle of torsion.   The axis of the movable coil should be North and South, so that it may not be effected by terrestial magnetism.

A large class of voltmeters which may be called *attraction* or *electro-magnetic* voltmeters depend on the principle that when a current flows through a coil of wire it creates a magnetic field which attracts a piece of soft iron towards the strongest portion of the field.

This attraction is, of course, independent of the direction of the current.

In a form of instrument manufactured by Hartman and Braun, a small soft iron coil is held by means of a spring just above the attracting coil, and the motion is multiplied by means of a lever arm moving over a graduated scale.

One form of the Ayrton and Perry voltmeter is similar to the above, except that the spring is placed within the coil and the arrangement of the indicating hand is somewhat different.

In Evershed's voltmeter, Fig. 68, a piece of soft iron, s s, a current flows through the coil it tends to rotate the soft iron core to a position in the axis of the coil.

In the magnetic vane voltmeter, the p. d. is measured by the repulsion exerted between a fixed and movable vane of soft iron placed in the field of a magnetizing coil, the action of the movable vane being opposed by a spring.

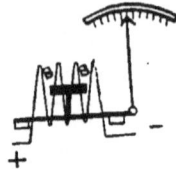

. FIG. 68.

Just what errors may be caused by hysteresis or residual magnetism in the above class of voltmeters it is difficult to say, but it is probable that most of them are fairly correct and well adapted to the class of measurements for which they are employed.

It should be understood that the instruments just described may be employed as voltmeters or ammeters, depending on whether they are given a high or a low resistance.

Caloric voltmeter. When a current flows through a wire, the wire is heated and expands. The amount of heating or expansion is proportional to the current and also to the p. d. across the ends of the wire.

In the "hot wire" voltmeter, the amount of this expansion is indicated by a pointer held in position by springs, the scale being graduated in volts. Fig. 69, shows the principle of construction of the instrument.

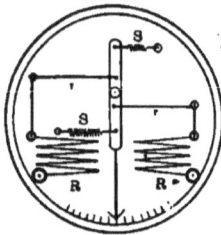

FIG. 69.

In the Cardew voltmeter as formerly manufactured, the expansion wire was placed in a long tube at the side of the indicating scale. In the more recent instruments, however, a circular case is employed and their appearance does not differ from the ordinary voltmeter.

The expansion of the wire is, of course, independent of the direction of the current. The readings are not effected by strong magnetic fields, and hence these instruments are very suitable for station work.

Some of these instruments, however, have a rather low resistance and require a considerable current to operate them, so that they are not always adapted to measurements when the resistance of the voltmeter is a factor in the determination.

Measurement of very high E. M. F.—The Thomson Electrostatic Voltmeter might be employed for the determination of E. M. F.'s considerably above 10,000 volts if the distance between the plates and the vane were made sufficiently great to prevent sparking and if the entire instrument were very carefully in-sulated. Heavier counter weights could be employed and each division deflection thus made to correspond to a greater p. d.

In the Absolute Electrometer and Kirchhoff's Balance, the attraction between a fixed and movable plate is measured and by means of suitable formula the e. m. f. calculated. Of course, the limit of measurement is determined by the striking distance of the spark and by the insulation of the apparatus.

Very great potential differences can be roughly calculated from the striking distance of the spark in air. It depends to a certain extent on the size and shape of the electrodes. The striking distance increases faster than the difference of poten-tial, and the curve indicating the ratios of striking distances to differences of potential is a parabola.

According to Lord Kelvin's measurements, the potential dif-ference required to produce a spark in air, between parallel plates, and of a given length, diminishes rapidly as the distance increases, approaching a limiting value of 30,000 volts per centi-metre, which may be assumed constant for distances greater than one centimetre.

For sparks not under two millimetres in length the volts ne-cessary to start a spark across a length of l centimetres may be approximately calculated by the formula—

$$V = 1,500 + 30,000 \, l.$$

Measurement of very low E. M. F.—A very sensitive galvano-meter will give a deflection of one scale division for a differ-ence of potential of .000001 volt across its terminals.

The galvanometer may be calibrated by means of a standard cell in series with a high resistance. Thus: suppose the resist-ance used is 100,000 ohms, galvanometer resistance 1000 ohms, and deflection 300 divisions, then 1 division deflection = 1,435

volts $\times \dfrac{1,000}{101,000 \times 300} = .000047$ volt.

With one form of the Weston voltmeter readings as low as $\frac{1}{300}$ volt can be obtained, and if the series resistance were cut out it is probable that it would indicate about $\frac{1}{3000}$ volt.

Quite small differences of potential may be observed by means of a capillary electrometer, which consists of a very finely drawn out glass tube containing mercury and 60 per cent sulphuric acid in contact with each other. A potential difference between them causes a change of capillary pressure at the point of con-tact, and hence a displacement, which for small potential differ-ence is proportional to the latter. This displacement may be accurately determined by means of a microscope.

CHAPTER XVI

Voltmeters are best calibrated by comparison with one or more standard cells by means of a potentiometer. The arrange-ment of the experi-ment is shown in the diagram Fig. 70.

Across the ends of the potentiometer RR', consisting of a slide coil bridge similar in construction to one of the forms previously described and of not less than 10,000 ohms resistance, is joined the voltmeter v.

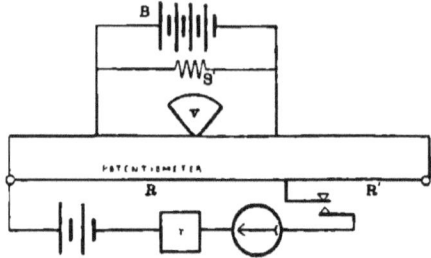

FIG. 70

In place of a slide coil bridge two rheostats may be employed for R and R' and the adjustments so made that the joint resist-tance is always 10,000 ohms. A battery of constant cells B, pre-ferably storage cells, with a shunt resistance s' is also connected to the terminals of the potentiometer. The E. M. F. of this bat-tery should be somewhat greater than the highest reading of the voltmeter. The standard cells E in series with a protecting re-sistance r, that may be short-circuited, and the galvanometer are joined to one terminal of the potentiometer and the contact key as slider. The positive poles of testing battery and standard cells should be connected to the zero terminal of the potentio-meter.

If it is desired to find the errors in a voltmeter scale already graduated the method is as follows: s' is adjusted until some convenient reading is obtained on the voltmeter, and then a balance is obtained on the potentiometer, shown by no deflec-tion of the galvanometer, the potential difference v across the ends of the potentiometer being calculated from the proportion.

R : R + R' :: E : V. Example, voltmeter reading = 10.7, E = 1.435 × 2, R = 2657, R + R' = 10,000 ; then 2657 : 10,000 :: 2.87 : V. V = 10.8 volts ; error of voltmeter reading is therefore — 0.1 volt and the correction + 0.1 volt.

By this means the errors in various parts of the scale are determined and a correction table or curve constructed.

When it is desired to graduate the scale or adjust the voltmeter by changing its resistance until the readings correspond with the scale, the following method is employed. Suppose a potential difference of exactly 10 volts is required across the voltmeter terminals, R is given such a value that R : R + R' :: E : 10, thus R : 10,000 :: 2.87 : 10, R = 2870. s' is then adjusted until the galvanometer gives no deflection. The potential difference across the potentiometer and consequently across the voltmeter must then be just 10 volts. The position taken by the indicating hand of the voltmeter is therefore marked 10, and so on for the other portions of the scale.

The advantage of using several standard cells is that a better average value for the E. M. F. is obtained and also that the readings on the potentiometer are larger, when the potential differences across the terminals are high, than if a single cell were employed. For very accurate work the standard cells should be placed in a water bath or an oil bath, and the correction, if any, for the temperature coefficient applied.

It should be understood that the above method is suitable for the calibration of standard laboratory voltmeters. After a voltmeter has once been accurately calibrated, other voltmeters may be checked by a direct comparison with it.

A voltmeter may also be checked by taking the reading across a known resistence R through which a known current C is flowing. The potential difference E across R may then be calculated from the formula $E = C R$. The difference between this and the voltmeter reading, of course, gives the correction. The arrangement is similar to that shown in Fig. 62 except that the voltmeter terminals are joined across R in place of the cell E, galvanometer, etc. By varying the current different readings may be obtained. For R, a standard ohm or some accurately measured resistance that will not be heated by the current should be employed. This resistance should be low compared to the voltmeter resistance. The current can be measured by means of a Thomson balance, a calibrated ammeter, or if it be constant, with a voltmeter.

Standards of E. M. F.—In the case of E. M. F., there is considerable difference in the values of the unit and of the standard.

The "absolute unit" of E. M. F. is the E. M. F. developed by a conductor when it cuts one "line of force" per second. The practical unit or volt is 10⁸ absolute units. The international standard of E. M. F. adopted is the Clark standard cell. Its E. M. F. is 1.434 true volts at 15°C. It consists of an anode of pure zinc in a concentrated solution of zinc-sulphate and a cathode of pure mercury in contact with a paste of pure mercurous sul phate. Precise directions are given for setting up these cells, and if the directions are followed they may be relied upon to give the E. M. F. stated above. The variations of the E. M. F. with temperature may be calculated with sufficient accuracy from the formula :

$$E = 1.434 \{1 - 0.00077 \; (t-15)\},$$

where t is the temperature of the cell in degrees centigrade.

Various other standard cells are made use of in practice. The Carhart-Clark cell employs a solution of zinc-sulphate saturated at 0°C. Its E. M. F. is 1.440 true volts at 15°C. and the temperature coefficient is approximately 0.00038 per degree c.

In the Weston standard cell, a cadmium anode is used immersed in sulphate of cadmium. The E. M. F. is 1.025 true volts and this value is practically constant at all ordinary temperatures.

Standard Clark-cells prepared according to the specifications are best checked by comparison with each other by the potentiometer method, care being taken that they are of the same temperature. If several are found to agree to the third decimal place, it may be taken as very certain that the E. M. F. is 1.434 true volts at 15°C. Other standard cells may then be compared with them by the same method.

The E. M. F. of standard cells may also be determined by connecting in series with a galvanometer and protecting resistance and joining the terminals of this branch circuit across a known resistance R through which a current from a constant battery is flowing. The arrangement of the experiment is shown in Fig. 62. The current is varied by means of the resistance s' until the potential difference across R is equal to the E. M. F. of the standard cell shown by no deflection of the galvanometer. The current can be measured accurately by means of a silver voltmeter or by use of a Thomson Balance. Then E = C R.

It may be well to state that the E. M F. differs according as it is expressed in true or international volts, legal volts, or B. A. volts. The ratios are the same as those existing between the the corresponding values of the ohm.

CHAPTER XVII.

CURRENT.

The presence of an electrical current is manifested by several accompanying phenomena and it is by the observation of the intensity of these phenomena that current strength is determined.

A conductor carrying a current is surrounded by a magnetic field, and the strength of this field may be measured by the deflective action on a magnetic needle, the attractive force exerted on a piece of soft iron, the attraction between two coils through which the current is flowing, etc.

Besides this magnetic field there is also an electrostatic field surrounding the conductor, in other words, there is a difference of potential all along a conductor through which a current is flowing. This potential difference can be determined and the resistance across which it is measured being known, the current may be calculated.

Whenever a current flows through a conductor heat is developed and this calorific effect is proportional to the square of the strength of the current.

If a current be made to flow through an electrolyte, chemical action takes place and this chemical action is also a measure of the current strength.

The methods and instruments for the measurement of current are quite numerous and, of course, vary according to the special cases required.

One method, however, is universally applicable. This is the measurement of the potential difference across a known resistance. For this determination any of the methods used for the measurement of P. D. can be employed.

It may, perhaps, be found more convenient at times to make use of some of the other methods; so it is well to briefly consider them.

DIRECT CURRENTS.

E. M. F. and Resistance.. { Direct Method.
Differential Method (Cardew's).
Bridge Method (Kempe's).

P. D. and Resistance..... { Direct Deflection Method.
Equilibrium Method.
Potentiometer Method.
Voltmeter Method.*
Galvanometer Method.*

Tangent Galvanometer.

Voltameters............. { Weight.
Volume.

Ammeter.

{ Also the methods given for Alternating
Currents.

E. M. F. and resistance, direct method.—To measure current by this method the deflection of a low resistance galvanometer is noted when placed in the circuit through which flows the current whose strength is to be determined. The galvanometer is then joined in series with a battery of known E. M. F. and a rheostat,

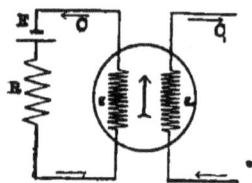

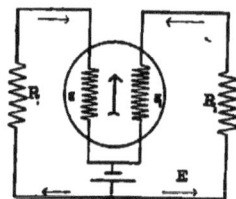

FIG. 71.   FIG. 72.

the resistance being adjusted until the deflection is equal to that obtained in the first case. Then the current may be cal-·culated from the formula

$$C = \frac{E}{R + G + r}.$$

Where G = resistance of galvanometer and r = resistance of battery. These latter may be neglected if small compared to R. This method is applicable to the measurement of comparatively weak currents.

Cardew's Differential Method.—For this method the galvanometer is wound with two coils $g\ g_1$, Fig. 71. Through the coil g_1, which is of low resistance, is passed the current to be measured. To the other coil g is joined a cell of known E. M. F. and a resistance R. This resistance is adjusted until there is no deflection, then

$$C_1 = \frac{E}{R + g + r} \times \frac{d}{d_1},$$

where d and d_1 are respectively the relative deflective effects of the coils g and g_1. The values of d and d_1 may be determined by joining up a battery and two resistances R_1 and R_2 in the manner shown in Fig. 72, and then adjusting until equilibrium is produced ; then $\dfrac{d}{d_1} = \dfrac{R_1 + g}{R_2 + g_1}$

Kempe's Bridge Method.—This method is a modification of the preceding one, and has the advantage that it does not require a special form of galvanometer.

In making the measurement the resistance R, Fig. 73, is adjusted until no deflection is observed, then

$$C_1 = \frac{E\,r}{r_1\,(R + r)},$$

where C_1 is the current to be measured and E the E. M. F. of the auxiliary cell whose resistance should be negligible compared to R. It is well to give the resistances r and r_1, a ratio of say 100 : 1.

$$\text{Example}: C_1 = \frac{1.08 \times 100}{1\,(4000 + 100)} = .026.$$

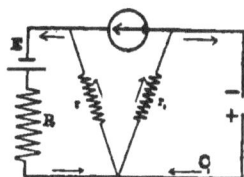

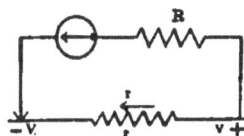

FIG. 73.   FIG. 74.

P. D. and Resistance.—The P. D. across a known resistance through which a current is flowing may be determined in a great variety of ways and from this the value of the current is readily calculated.

If a galvanometer in series with a high resistance R, Fig. 74, be joined across a low resistance r through which a current is flowing and the deflection d_1 observed, and then if the galvanometer and high resistance be joined up with a cell E of known E. M. F. and the deflection d read, the P. D. across r, $V - V_1$, is obtained from the proportion $V - V_1 : E :: d_1 : d$.

A galvanometer and cell E of a known E. M. F. can be connected across r and this resistance varied until equilibrium is obtained, then $E = V - V_1$. It is evident that such a method would only be applicable in special cases. A potentiometer might be employed to measure the P. D. where considerable accuracy was required. The P. D. could also be determined by charging a condenser across r.

One of the most convenient and accurate methods of measuring current is to employ a Weston voltmeter with a low reading
scale across a shunt R through which the
current is flowing, Fig. 75. Then if E be
the voltmeter reading, $C = \dfrac{E}{R}$.

It is well to have a set of shunts of
say the following values, .001, .01, 0.1, 1.
ohm respectively. The low resistances
to be used for heavy currents and the
higher resistances for light currents.

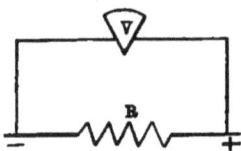

FIG. 75.

Where very small currents are to be measured, a high resistance galvanometer should be joined across a 1-ohm shunt,
through which the current flows, and the deflection observed.
The value of the galvanometer deflection per scale division in
microvolts can afterwards be obtained by means of a standard
cell and high resistance. Since the best galvanometers have a
sensitiveness of 1 microvolt, it is possible to measure current
to the one-millionth of an ampere with a one-ohm shunt.

The strength of current that can be determined by this
method, when voltmeters and shunts of very low resistance are
employed, is, of course, practically unlimited.

Tangent Galvanometer.—If the diameter of the coils of a galvanometer is great, compared to the length of the needle, the
tangent of the angle of deflection is proportional to the current
flowing through the instrument.

The same effect may be obtained by placing the coils sym-
metrically each side of the needle. For strong currents, a single turn of very thick copper wire should be employed.

The "constant" of the galvanometer, or the amount of
current necessary to produce 1° deflection, depends on the radius of the coils, number of turns, and strength of field. The
value of this constant can be obtained by observing the deflection with a battery of known E. M. F. and a known resistance in circuit, or by comparison with a voltmeter, calibrated
ammeter, standard voltmeter across shunt, etc.

The tangent galvanometer gives a ready means for the comparison of current strengths, but the use of the ammeter or
shunted voltmeter, for most work, is much to be preferred.

Voltameters.—According to the resolutions adopted by the
Electrical Congress of 1893, the international ampere is considered as represented sufficiently well for practical use by the
unvarying current which, when passed through a solution of
nitrate of silver in water, and in accordance with the specifica-

tions given, deposits silver at the rate of 0.001118 of a gramme per second.

A neutral solution of pure silver nitrate, containing about 15 per cent. by weight of the nitrate and 85 per cent. of water should be employed. A current strength of about one ampere should be used and the specifications strictly followed if the most accurate results are desired.

The value of the current can be obtained from the formula, C = Weight of silver deposited $\div$ (0.001118 × Time in seconds).

For approximate work a voltameter consisting of copper in a solution of copper sulphate may be employed. An ampere will then deposit 0.0003281 of a gramme of copper per second.

The objection to the use of voltameters is that the conditions of experiment, such as the strength of the current, size of the electrodes, strength of the solution, etc., may effect the accuracy of the results.

If a current flows through water to which sulphuric acid has been added, the water is decomposed and hydrogen and oxygen liberated. These gases can be collected in graduated tubes and their volume measured from which the current strength can be calculated. It is better in practice to measure only the hydrogen, for a portion of the oxygen is condensed to ozone, and it is also slightly soluble in the solution.

An ampere liberates .1155 cubic centimeters of hydrogen per second, if the volume be taken at the barometric pressure of 760 mm. and the temperature of 0° C. The volume observed must therefore be reduced to these standard conditions by the use of suitable formulæ. Then the current in amperes =Volume in C. C. $\div$ (1155 × Time in seconds).

Ammeters.—The following classification indicates a few of the more important forms of instruments whose scale readings give current strength directly.

AMMETERS.
{
 Weston.
 Ayrton and Perry.
 Hartman and Braun.
 Evershed's.
 Schuckert's
 Magnetic Vane.
}

The Weston ammeter is almost precisely similar in construction to the Weston voltmeter, shown by Fig. 64. The movable coil, however, instead of being connected in series with a high resistance, is joined in parallel with a low resistance. This shunt in some of the instruments consists of a number of copper wires wound in parallel about the magnet and has about .002 ohm resistance.

It is hardly necessary to say that the portable standard form of these ammeters are extremely reliable and accurate instruments.

One form of the Ayrton and Perry ammeter is shown in Fig. 76. A short magnetic needle is placed between the pole pieces of a powerful permanent magnet which controls its direction and renders it independent of the earth's magnetism. When a current flows through the solenoid, it tends to rotate the needle toward the axis of the coil. This coil is of very low resistance and consists of but few turns of copper wire. By the proper shaping of the pole pieces, needle, and coil, the angular deflections are made proportional to the strength of the current.

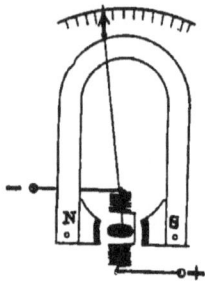

FIG. 76.

An ammeter largely manufactured by Hartman and Braun, of Frankfort, and used to a considerable extent abroad, is shown in Fig. 77. A very small tube, made of soft, thin sheet iron, is attached to a spring and lever arm, and placed near the end of an attracting solenoid consisting of a few turns of stout copper wire. By correctly shaping the tube, the deflections can be made proportional to the strength of current. It is claimed in the most recent instruments, that the effects of residual magnetism and hysteresis are practically eliminated. This instrument in somewhat simpler form is known as the Kohlrausch ammeter.

FIG. 77.

The Evershed instrument, known as the "Gravity" ammeter, consists of a magnetizing coil, placed with its axis horizontal. In the older form of these instruments, a small cylindrical piece of soft iron fixed to a spindle pivoted at its extremities, is placed within a coil and counter-weighted to maintain in certain position. When a current flows through the coil, the iron is rotated toward a position between the ends of two plates of iron also within the coil, but not shown in the diagram.

For very heavy currents, the magnetizing coil consists of a massive tubular casting of copper divided by saw-cuts to form a "coil."

In the more recent form of this instrument, two curved pieces of iron are placed within the coil. The outer is fixed and concentric with the inner one, which is mounted on the counter-weighted spindle. When a current flows through the coil, the movable piece of iron is urged round towards the position where the two pieces would form approximately a complete tube of iron.

This same construction is also used for the Evershed voltmeters.

A possible source of error in the above described ammeters is the retentivity of the iron. It is, therefore, well to test the readings with an increasing and decreasing current.

In the Schuckert ammeter, an index is pivoted in the axis of the magnetizing coil, and carries a light strip of soft iron. Another strip is fixed with the coil. When a current flows through the coil, these strips become magnetized and repel one another. The controlling force is gravity.

The principle of the "Magnetic Vane" ammeter is similar to the above, the motion of the movable vane being opposed by a spring.

With regard to the various forms of ammeters just described, depending on the magnetic effect of a coil on iron, it is not easy to say just how reliable or unreliable any particular instrument under varying conditions may be. Their chief advantage is, that they can be made more sensitive for a certain portion of the scale, that is, for a given strength of current, and that they may be employed for alternating as well as direct currents.

Whenever possible, however, the use of a Weston ammeter is to be recommended, or, better still, a standard Weston voltmeter, across a shunt, for in this case, by varying the resistance of the shunt, the range of measurement is unlimited.

ALTERNATING CURRENTS.
{
  Dynamometer.
  Current Balance.
  "Attraction" or Electro-Magnetic Ammeters.
  P. D. and Shunt.
  Calorimetric Method.
}

It should be understood that the methods and instruments described for alternating currents are also suitable for the measurements of direct currents.

The dynamometer and current balance might have been classified with ammeters, since they are current measuring instruments, but on account of their importance it is perhaps better to treat them separately.

Dynamometer.—The principle of the electrodynamometer has been previously explained, and is shown in Fig. 67. For the measurement of current, the coils should be of very low resistance.

In the Siemens dynamometer, much used for the measurement of strong currents, whether direct or alternating, one coil is fixed permanently, whilst the other coil, of one or two turns, dipping with its ends in mercury cups, is hung at right angles, and controlled by a special spring below a torsion head. When a current passes, the movable coil tends to turn parallel to the fixed coil, but is prevented; the torsion index being turned until the twist on the spring balances the torque. The angle through which the index has had to be turned is proportional to the square of the current strength.

The axis of the movable coil should be in the line of the magnetic meridian, and the coils should be accurately perpendicular to each other.

Where current strength is determined by the deflections of a dynamometer, the mean current strength of an alternating current is $\frac{9}{10}$ of the strength of the continuous current, which would give the same deflection.

Current Balance.—The principle of the Thomson current balance is indicated by Fig. 78. There are four fixed coils,

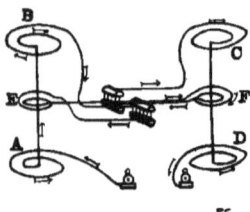

FIG. 78.

A, B, C, D, between which is suspended, by a flexible metal ligament of fine wires, at the ends of a light beam, a pair of movable coils, E and F. The current flows in such directions through the whole six, that the beam tends to rise at F, and sink at E. The beam carries a small pan at the end F, and a light arm along which a sliding weight can be moved to balance the torque due to the current. The current is proportional to the square-root of this torque, the force being proportional to the product of the current in the fixed and movable coils, as in all electro-dynamometers. The current balance is in fact a current weighing dynamometer.

A complete range of these instruments has been designed, reading from .01 ampere to 2,500 amperes.

The Thomson balance forms a most reliable standard for the measurement of current and the calibration of other instruments.

When these instruments are made so as to measure alternating as well as continuous currents, the current is carried by a twisted rope of copper wires, each of which is insulated. The object of this arrangement is to prevent inductive action.

" *Attraction*" or *Electro-Magnetic Ammeters.*—Alternating currents may be measured with more or less accuracy by the various forms of these ammeters, such as the Evershed, Schuckert, etc., previously described.

When such an electromagnetic ammeter is employed for the measurement of alternating currents, the general tendency is for its readings to be lower than the correct value, if it is calibrated to be correct for direct currents, chiefly on account of the eddy currents which are set up in the framework and metal parts of the instrument. It is found that the Evershed ammeters indicate about two per cent. lower than the true value of such an alternating current. This error, however, is corrected by permanently shunting the main ammeter coil by a smaller coil of copper wire which is overwound with thin iron wire, in order to raise its self-induction to the desired value.

If there be any hysteresis or retentivity in the iron used in this class of ammeters, the error caused by it may be considerable.

P. D. and Shunt.—An alternating current may be conveniently and accurately determined, if the potential difference across a shunt of known resistance be measured by means of a Weston alternating current voltmeter or some form of electrometer. Since these instruments are not very sensitive for small potential differences, and the resistance of the shunt must necessarily be low, the above applies especially to strong currents.

Calorimetric Method.—The heat units, or calories developed by a current in a given time, is equal to $.24\ C^2\ R\ T$, where C is the current in amperes, R the resistance, and T the time in seconds. Therefore, if this amount of heat be determined by means of a calorimeter, the current strength can be calculated. The method, however, is not a very practical one.

If the expansion of a wire were used to indicate the current, as in the case of the Cardew voltmeter, the resistance would be too high to introduce into the circuit.

$$\begin{cases} \textit{Very High Currents} \\ \text{and} \\ \textit{Very Low Currents.} \end{cases}$$

It has been stated that the measurement of potential difference across a shunt of known resistance is a universal method for the measurement of current, and it is especially desirable for the determination of very strong or very weak currents.

Suppose the standard form of Weston voltmeter be employed, with which readings can be made from $\frac{1}{300}$ volt to 150 volts, and that a shunt of .0001 ohm be used. Then the range of measurement would be from 33 amperes to 1,500,000 amperes.

Now, the resistance of such a current could be very accurately measured by means of the double bridge, the best plan being to determine the resistance between two marks on a heavy bar of copper. The leads from the voltmeter should then be connected

to knife edges resting upon these marks. Since the resistance
of the Weston voltmeter, even when the low reading scale is
used, is about 500 ohms, the resistance of the leads and contacts
would be entirely negligible compared to it.

A mistake probably often made when very heavy currents are
to be measured, is that of shunting a low reading ammeter.
Here the case is entirely different, for then the contact resistances are added on to the low resistance of the ammeter and
may produce a considerable error, even though the shunt has
been most accurately adjusted.

Since a sensitive high resistance galvanometer will indicate a
micro-volt, if the galvanometer be shunted across one ohm, it is
then possible to measure current to one millionth of an ampere.

Calibration of Ammeters.—The best method of calibrating an
ammeter is to compare the readings with those of a standard
Weston voltmeter shunted across a known resistance. The arrangement is shown in Fig. 79, R being the resistance, and r an
adjustable resistance to vary the current.

The correct strength of current is, of course, given by
the voltmeter reading divided by
the resistance of the shunt. In place
of the voltmeter, the potentiometer
can be used, according to the method
given for calibrating a voltmeter.

Since the voltmeter may be compared to the Clark cell by means of
the potentiometer, and the shunt resistance to the standard ohm, the standards of E. M. F. and resistance become
also the standards for current.

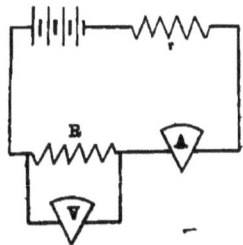

FIG. 79.

By means of the Thomson balance, an ammeter can be very
accurately calibrated.

The ammeter can also be checked by comparison with the
voltmeter.

When it is desired to compare a low reading ammeter that
has been calibrated with a high read-
ing ammeter, the arrangement shown
in Fig. 80 can be employed. If the
resistance of the shunt r is equal to $\frac{1}{99}$
of the resistance of ammeter 1 + R,
then ammeter 1 will only receive .01 of
the entire current that flows through
ammeter 2. By this means an ammeter only reading to 15 amperes could
be compared with one reading to 1,500 amperes.

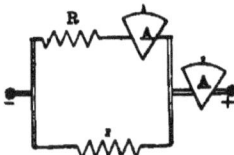

FIG. 80.

Absolute Determination of Current.—Current strength can be measured in absolute units from the deflections of a tangent galvanometer, if its radius, number of turns, and strength of surrounding field be known. The equation is :

$$C = \frac{r}{2\,n\,\pi} \times H \times \tan a \,,$$

where r is the radius of the galvanometer coils in centimetres, n the number of turns, H the strength of field, and a the deflection.

Considerable care should be used in this determination, if accurate results are desired.

This method is interesting on account of its employment in the determination of the value of the standard ohm, and also forms an additional check on the other methods of current measurement.

CHAPTER XVIII.

$$\text{Energy.}\begin{cases} Wattmeter. \begin{cases} \text{Weston's.} \\ \text{Siemens'.} \end{cases} \\ Voltmeter\ and\ Ammeter. \end{cases}$$

The amount of electrical power consumed by lamps, motors, etc., can be directly measured by means of instruments known as voltmeters.

The unit of electrical energy is the watt, or kilowatt (= 1,000 watts), and a horse-power is equivalent to about 746 watts.

Most of these wattmeters are modifications of Weber's dynamometer, in which a fixed coil produces a field, and tends to turn a movable coil. One of the best known wattmeters is Siemens' dynamometer, in which one coil is wound with fine wire and is put in shunt to the part of the circuit in which the power is to be measured, and a thick wire coil which is joined in series. The force is then proportional to the product of the currents in the two coils, that is, to the product of the potential difference and current, or to the power. In the Weston wattmeter, the motion of the movable coil is opposed by a spring in a manner similar to that used in the voltmeter and ammeter.

The resistance of the pressure or shunt coil should be as high as possible, since the current that it takes also passes through the series coil, and may thus cause a considerable error. As the pressure coil generally takes more power than the current coil, it is best to put it in shunt to the current coil in addition to the lamp or other device across which the power is to be determined. Some wattmeters are compensated for this error.

Electrical energy can be very readily determined by means of the voltmeter and ammeter. The voltmeter is used in shunt and the ammeter in series (Fig. 81), and the power is then obtained by multiplying the potential difference

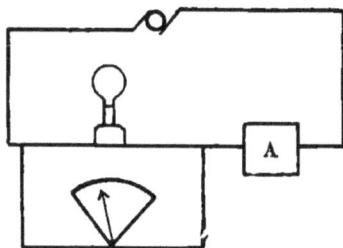

FIG. 81.

100

indicated by the voltmeter by the reading of the ammeter. If a Weston standard voltmeter be employed, the error caused by the current taken by the voltmeter is very small. This error may be still further reduced by placing the voltmeter also in shunt to the ammeter.

$$\text{QUANTITY.} \begin{cases} \textit{Voltameter.} \quad \text{(Edison Meter.)} \\ \textit{``Meters.''} \\ \textit{Ballistic Galvanometer.} \end{cases}$$

The amount of electrolytic action in any voltameter is proportional to the strength of current and the time ; that is to say, it is proportional to the quantity of electricity. The unit of measurement is the ampere second or coulomb. The practical unit is the ampere hour.

The voltameter generally used in practice is the Edison "chemical" meter. It consists of two jars of zinc sulphate with zinc electrodes so connected across a shunt that they receive, say, $\frac{1}{1000}$ of the entire current. The resistance of an electrolyte decreases with a rise in temperature. To compensate for this error, copper wires are joined in series with the cells. Two cells are used for greater accuracy, the amount the electrodes lose in weight in each being determined. These two results should, of course, check each other. The coulomb deposits 0.33696 milligramme of zinc, and the ampere-hour 1,213 milligrammes. The arrangement of the Edison meter is shown in Fig. 82.

It is extremely difficult to measure satisfactorily electrical quantity on a commercial scale. A number of instruments have been devised for this purpose, and they are known under the general name of "meters." Prob-

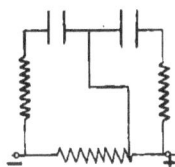

FIG. 82.

ably one of the best of these meters is the Thomson-Houston recording wattmeter. It consists essentially of two thick wire coils placed in series in the circuit, and a thin wire coil placed in shunt around the circuit whose power is to be measured. The shunt coil is mounted on an axle carrying a copper disk moving between the poles of permanent magnets. Under these conditions, the rate of rotation produced in the movable coil is proportional to the energy consumed in the main circuit. The number of revolutions is recorded by clockwork and the instrument is graduated to indicate watt-hours, etc.

In the Forbes' meter, the current passes through a number of fine wires placed in parallel. These wires becoming heated, produce a rising current of warm air, and this rotates a spindle carrying mica vanes.

The Ferranti meter consists of a vessel containing mercury, above which is placed a solenoid. The current is led to the mercury at the centre of the vessel and leaves it at the circumference, then passing through the magnetizing solenoid, the mercury is urged to move in a direction at right angles to that in which the current is flowing through it, and also at right angles to the lines of force of the field. This, of course, produces rotation. The amount of rotation is measured by means of a float geared to the proper indicating device.

The Aron meter consists of two clocks geared differently. The pendulum of one clock carries a permanent magnet. Beneath this is placed a solenoid, through which flows the main current. When both pendulums oscillate at the same rate, no movement of the indicating pointers takes place, but they begin to indicate if one of the pendulums is accelerated. This acceleration is proportional to the strength of the current flowing through the solenoid. They can be adjusted to indicate amperehours.

In order to make this or any similar meter show the energy consumed, or watt-hours, it is necessary to multiply the amperehours by the pressure at which the current is supplied. Therefore the accuracy of the result depends upon the constancy of the pressure as well as the accuracy of the instrument.

The deflections of a ballistic galvanometer are proportional to the quantity of electricity passing through the galvanometer, if the discharge occupy a very short time compared to the time of vibration of the galvanometer needle. The application of this fact, however, is in the absolute determination of capacity and inductance.

CHAPTER XIX.

CAPACITY.

DEFLECTION METHODS.
$\left\{\begin{array}{l} \textit{Direct Deflection.*} \\ \textit{Direct Charge.} \\[1em] \textit{Loss of Charge.} \left\{\begin{array}{l} \text{Discharge.} \\ \text{Deflection.} \end{array}\right. \end{array}\right.$

ZERO METHODS .. $\left\{\begin{array}{l} \textit{Bridge Method.} \\ \textit{Potentiometer Method} \text{ (Mixtures.)*} \end{array}\right.$

ABSOLUTE DETERMINATION (Ballistic Galvanometer.)

Electrostatic capacity may be defined as the ratio of the quantity of any electrical charge to the E. M. F. producing that charge, or $F = \dfrac{Q}{E}$. Scientifically speaking, it is the ratio of dielectric strain to dielectric stress, the term " quantity " of electricity being used only as a matter of convenience. The unit of capacity, or the farad (F), is such a capacity that the unit quantity, one coulomb, is obtained under the pressure of one volt. This capacity is far too large tor ordinary measurements, so the practical unit employed is a millionth of this, or the micro-farad.

The accurate determination of capacity in many cases is impossible, since most condensers, to a certain extent at least, and practically all cables exhibit the phenomena of absorption and residual charge. Therefore, when the capacity is stated, all the conditions of measurements should be given.

Direct Deflection.—In this method, a standard condenser, F, is charged by a battery, B, Fig. 83, and then dicharged through a high resistance galvanometer, and the deflection d observed. The unknown condenser, F_2, is then substituted, and the deflection d_2 noted. Then $F_1 : F_2 : : d_1 : d_2$.

Some uniform time of charge, such as five seconds, should be adopted. Several observations should be taken in each case, and the mean used in the calculation. The method is suitable and convenient where only approximate results are desired.

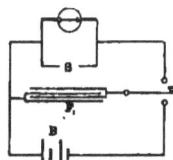

FIG. 83.

The absorption of various condensers may be studied by this method by observing the deflection after charging for different lengths of time, such as 1 second, 30 seconds, 1 minute, etc. The residual charge can be determined by discharging, insulating for one minute, and discharging again, insulating for another minute, etc.

It is important in the above method that there be no self-induction in any portion of the circuit, or in the galvanometer shunt, if it be employed, for, of course, this would change the value of the deflections and thus cause an additional error in the measurement.

Divided Charge.—The connections for this method are shown in Fig. 84. The standard condenser, F, is charged by closing the battery key, k. It is then discharged, and the deflection, d, noted. It is again charged, the key k is opened and the key K depressed for a few seconds, by this means allowing the charge to divide between the two condensers, F_2, being the unknown condenser or cable. The standard condenser is then once more discharged. Call this deflection d_2, then

FIG. 84.

$F_1 : F_2 : :d_2 : d_1 - d_2$, for the quantity of charge in each condenser is proportional to the capacity. This method is said to be very accurate for the measurement of the capacity of long cables.

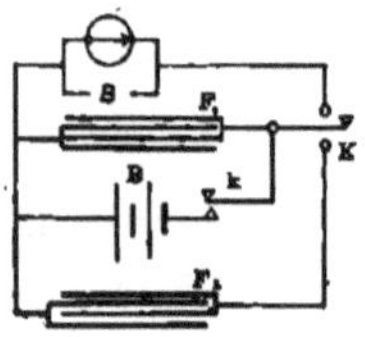

Loss of Charge—Discharge.—The capacity of a condenser can be calculated from the formula

$$F = \frac{T}{2.303 \ R \ (\log d_1 - \log d_2)}$$

when d is the discharge deflection obtained immediately after charging, d_2 the deflection after charging, and then insulating for T seconds, R the resistance between the poles of the condenser (if this be expressed in megohms, the capacity will be obtained in micro-farads), and 2.303 the modulus to convert the ordinary or Brigg's logarithms to natural logarithms. The connections are the same as Fig. 83. If a mica condenser be used, a resistance of several megohms may be placed between the poles. To measure the capacity of cables by this method, the insulation must be determined and this value substituted for R. Since the insulation is such a variable quantity, the above method is only very approximate.

Deflection.—A modification of the method just described is shown in Fig. 85.

The steady deflection is first observed with the key closed, d; it is then noted after τ seconds d_2, and the capacity calulated from the formula given above. The resistance, R, should · be great enough, several megohms, so that the charge will not be lost too rapidly. If the resistance of the condenser be low or if a cable is used, and if this resistance be callled r, then the

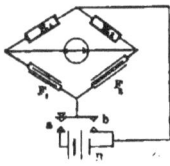

FIG. 85.

value of the resistance to be used in the above equation is

$$\frac{Rr}{R + r}.$$

Bridge Method.—Zero methods have the advantage that the errors due to reading the galvanometer deflections are avoided, and that the effects due to induction may be partially, if not entirely, eliminated.

The Bridge method is applicable to ordinary condenser work and to short lengths of cable, but is not suitable for great

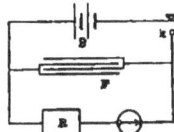

FIG. 86.

lengths of cable, on account of the influence of inductive retardation. The connections for the measurement are shown in Fig. 86. The method is very similar to the Wheatstone bridge. When the resistances R_1 R_2, which should be high, are so adjusted that there is no deflection of the galvanometer on making contact at a or b, then $R_1 : R_2 : : F_2 : F_1$. That

is, the capacities are inversely proportional to the resistances. During the adjustment of R_1 R_2, contact should be made at the point b, in order that the condensers are kept discharged. If the insulation of the condensers be not good, of course, an error may be caused by the current flowing through the condensers.

Method of Mixtures (Thomson's Method).—This method may be considered the standard for cable work, and is also very suitable when the most accurate comparison of condensers is desired. The method depends on the principle that the "quantity" of electricity in a condenser is equal to its capacity, multiplied by the P. D. of the charge, or $Q = F E$. If, then, two condensers $F_1 F_2$ have the same charge, $Q = F_1 E_1 = F_2 E_2$, or $F_1 : F_2 : : E_2 : E_1$. In this method the ratios of $E_1 E_2$ are the same as the resistances $R_1 R_2$; hence, $F_1 : F_2 : R_2 : R_1$. The arrangement for this measurement is shown by the diagram, Fig. 87.

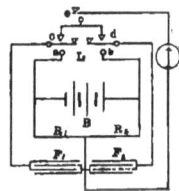

FIG. 87.

The rheostat R_1 R_2 should be of high resistance. It is convenient to use in place of them one of the "potentiometers" or slide coil bridges previously described. It is best to employ a special key, known as the Lambert capacity key, indicated in the diagram by I..

The manipulation is as follows : Contact is made at the points $a\,b$, and the condensers are thus charged across R_1 and R_2. Contact is then made at the points $c\,d$, and by this means the charges of the condensers are allowed to mix. Finally, contact is made at e, and the galvanometer, being thus placed in circuit with the condensers, is deflected, if the charges are unequal. The adjustment of R_1 R_2 is repeated until the galvanometer shows no deflection. Some standard time of charging should be employed, say, ten seconds, and the charges should be allowed to mix ten seconds. For long cables, a five minute charge is recommended and a time of mixture of ten seconds.

The values of F_1 F_2 should not be very unequal—that is, F_1 should not be much less than $\frac{1}{4}$ of F_2, for if the capacities are very different, the potential of one charge may be so much higher than that of the other that an error may be caused by absorption.

Absolute Determination.—The "quantity" of electricity which discharged through a ballistic galvanometer will produce a given deflection is expressed by the equation

$$Q = \frac{E}{R\,d_1} \times d_2 \times \frac{T}{2\,\pi} \times \left(1 + \frac{\lambda}{2}\right).$$

In the above equation $\frac{R\,d_1}{E_1}$ is the "constant" of the galvanometer, that is, if a potential difference E_1 be used through a resistance R, a steady deflection d_1 is obtained. d_2 is the throw of the galvanometer produced by the quantity Q, T the time in seconds of a complete or double vibration of the galvanometer on open circuit, and λ is the logarithmic decrement.

If a condenser of capacity F be charged by a potential difference E_2, then since

$$Q = F\,E_2, \qquad F = \frac{Q}{E_2},$$

consequently

$$F = \frac{d_2\,T\left(1 + \frac{\lambda}{2}\right) E_1}{R\,d_1\,2\,\pi\,E_2}.$$

In a ballistic galvanometer, the time of vibration of the moving system should be slow, the moment of inertia large, and the decrement or damping but slight. These conditions are fulfilled by several forms of galvanometer. One in which bell magnets are employed, shown in Fig. 7, and also the special forms of the D'Arsonval and the Ayrton and Mather galvanometer previously described. Either of the two latter galvanometers is much to be preferred for practical work over the first form, in which the magnetic system is movable.

To observe the time of vibration T, the galvanometer is given a vibration of 200 to 300 scale divisions and time of, say, 10 or 20 vibrations determined, the mean of several sets of observation should be taken. If a galvanometer with movable magnetic

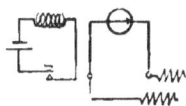

FIG. 88.

system is employed, the deflections are controlled by means of a "check coil," that is a solenoid in series with cell placed near the galvanometer, Fig. 88. If the D'Arsonval form of galvanometer is used, a cell and key may be placed in series with the galvanometer, or the galvanometer may be brought to rest by short circuiting.

The decrement is the ratio of the amplitude of any vibration to that of the next succeeding vibration. To obtain this, the tenth vibration after the first should be observed. Suppose this ratio is 1:40; then the decrement equals 1.04. If the decrement is small, such as the above example, then it is sufficiently accurate to call the factor $\left(1 + \dfrac{\lambda}{2}\right)$ equal to 1.02. Actually λ is equal to the log. of the decrement $\times$ 2.303.

To determine the "constant" of the galvanometer, the deflection d_1 is observed when a constant cell, such as a Daniell cell, or preferably a storage cell, is used. The galvanometer may be shunted, a high resistance placed in series with, or better still, the cell can be shunted. This last arrangement is shown in Fig. 89. In this case R is the resistance of the galvanometer and E_1 is equal to $E \times .004$.

Then condenser F is charged by the same cell, shunted if need be, and the deflection d_2 observed. The arrangement is shown by Fig. 90. Here E_2 is equal to $E \times .700$. Thus the E. M. F. of the cell need not be known, since it cancels the equation given above.

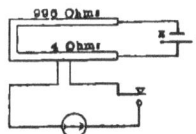

FIG. 89.

This method of measuring capacity is of considerable importance, for by it a standard condenser may be accurately calibrated. Of course, after the capacity of a standard is once accurately known, other standard condensers can be compared to it.

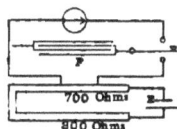

FIG. 90.

CHAPTER XX.

INDUCTANCE.

By the term "inductance" is meant the coefficient of self-induction.

When a current flows through a circuit, a magnetic field is established about the conductor carrying the current.

If the strength of the current rises, the strength of the field also varies. This has the effect of producing or withdrawing "lines of force," and if these cut adjacent wires in the circuit, an E. M. F. is developed in a direction opposite to that in which the current is flowing.

The unit of the inductance (L) is the henry or such an inductance that if the current varies one ampere per second, a counter E. M. F. of one volt is developed. From a consideration of the absolute system of units and dimensional formulæ, this has also been known as the "secohm" or "quadrant."

This coefficient may be determined by several methods.

Bridge Method.—This method requires a ballistic galvanometer. The coil s, whose inductance is required, is placed in the arm, *c d*, of a P. O. bridge, Fig. 91. The bridge coils, A, B, are made equal to each other, and as nearly equal to s as possible. An extra rheostat, R_2, is multiplied with R_1; by this means, an exceedingly fine adjustment can be obtained. All the resistances, except s, should be non-inductive.

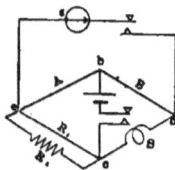
FIG. 91.

The resistance in the arm *e c* is adjusted until on closing, first the battery key and then the galvanometer key, no deflection is

observed. The galvanometer key is then first closed, and afterwards the battery key and the throw of the galvanometer, d_2
caused by the inductance of s obtained. The resistance in the
arm $e\ c$ is changed a small amount by altering the resistance in R_2.
Call this change of resistance in the arm $e\ e$ equal to r. The battery key is then closed, and the steady deflection, d_1 obtained on
closing the galvanometer key observed. The inductance is then
obtained by the equation

$$L = \frac{r\,d_2}{d_1} \times \frac{T}{2\,\pi} \times \left(1 + \frac{\lambda}{2}\right)$$

where T is the time in seconds of a complete or double vibration of the galvanometer, and λ the "logarithmic decrement."
These latter constants should be determined in a similar manner to that given for the absolute measurement of capacity.

The complete equation requires in place of $\frac{d_2}{d_1}$ the expression

$\frac{2\ \sin.\ \frac{1}{2}\ d_2}{\tan.\ d_1}$, but since the angle corresponding to d_2 is usually
small, the deflections as directly read off by means of a lamp
and scale, or telescope and scale, give the result with sufficient
accuracy for ordinary measurements.

The current, instead of being broken or made, may be reversed. In that case the value of d_2 is doubled, and by this
means a more accurate reading obtained.

Comparison with Standard.—If a coil (s_1) has been standardized
by the above method, then another inductance (s_2) may be compared to it without the use of a ballistic
galvanometer. The arrangement is shown
in Fig. 92.

The resistances B and R_2 are given some
constant value, such as 1,000 ohms each, and
are kept fixed. Then by the proper manipulation, A and R_1 may be so adjusted that no

FIG. 92.

deflection is obtained either for permanent currents or induction.
When this is the case, $s_1 : s_2 :: A : B$.

Secohmmeter Method.—If a "secohmmeter" or automatic in-
terrupter and commutator be employed in the battery circuit,
then the effect due to induction will be a steady deflection in
place of a throw.

The connections are the same as those shown in Fig. 91. The
resistance of A should be made equal to B, and the resistance of
the arm $e\ c$ adjusted to no deflection for steady currents. The
current is then interrupted by means of the secohmmeter, and
the resistance of the arm $e\ c$ changed by such an amount, r, that

no deflection is observed. Then, $L = \dfrac{r}{P}$, where P is the number of interruptions of the current per second.

Any ordinary sensitive galvanometer will, of course, answer for this method.

Adjustable standards of inductance are now made in the form of boxes of coils of different values, and also two coils in series that may be placed at different angles to each other, and the inductance in milli-henrys read off by means of a pointer and scale (Fig. 93). When these standards are at hand, the unknown inductance, s_2, is placed in one arm of a Wheatstone bridge, the standards, s_1, in the other, and A is made equal to B. The connections are indicated by Fig. 92. Then R_1 or

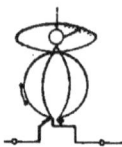
FIG. 93.

R_2 and s_1 are so adjusted that no deflection is obtained for either permanent or interrupted currents. When this adjustment is obtained, $s_1 = s_2$.

Condenser Method.—Inductance may also be compared to a capacity by the following method. The coil or electro-magnet, s, whose inductance is required, is joined in one arm of a Wheatstone Bridge, Fig. 94. In series with s is a resistance, r_1, call the resistance of s equal to r_2. Adjustment is made so that no deflection is obtained with permanent currents. The galvanometer key is then closed and afterwards the battery key and the throw of the galvanometer d_1 ob-

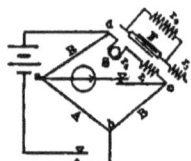
FIG. 94.

served. The coil s is then removed and a shunted condenser, F, is substituted in the arm $d\,c$. Balance for steady currents is again obtained, and the throw of the galvanometer on closing the battery key observed. Call this deflection d_2. Then if the adjustments have been so made that $r_1 + r_2 = r_3 + r_4$,

$$L = F\,(r_4)^2\,\frac{d_1}{d_2}.$$

A modification of the above method is shown in Fig. 95. s is the inductance to be measured and F a condenser shunted by a resistance, R. A balance for permanent currents should be obtained, and then the deflection d_1, on making the circuit, with the key f open, observed. Afterwards the deflection on making circuit with the key f closed, d_2 is obtained. Then,

$$L = F\,R^2\,\frac{d_1}{d_1 - d_2}.$$

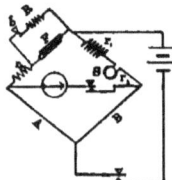
FIG. 95.

If the adjustments are so made that there is no

deflection in either case, $L = F R^2$. Or deflections may be obtained in opposite directions, and the value of F corresponding to no deflection interpolated.

Calculation.—The inductance of coils of known dimensions can be approximately calculated in some cases. Thus, in the case of a long uniform solenoid of length l centimetres, containing n turns of wire, the average radius of the turns being r,

$$L = \frac{4\,n^2\,\pi^2\,r^2}{l}$$

(approximately).

Impedance.—Impedance is the opposition to the flow of an alternating current. The reactance is equal to the inductance, L, multiplied by the period of alternation. The relation of these quantities to resistance is shown by Fig. 96. Thus,

$$\text{Impedance} = \sqrt{R^2 + p^2 L^2},$$

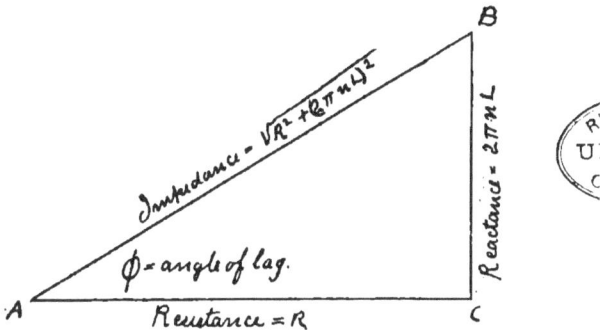

FIG. 96.

and therefore the average value of the current is given by the equation

$$C = \frac{E}{\sqrt{R^2 + p^2 L^2}}.$$

If there be also capacity in an alternating current circuit, a reactance is produced in a direction opposite to that given by inductance. It may be indicated by $-\dfrac{1}{p F}$, and the resultant reactance would therefore be equal to $p L - \dfrac{1}{p F}$. The current is then given by the equation,

$$C = \frac{E}{\sqrt{R^2 + \left(p L - \dfrac{1}{p F}\right)^2}}.$$

Capacity and inductance may be used to neutralize each other. If $L = \dfrac{1}{p^2 F}$, they exactly balance, and the circuit becomes non-inductive.

CHAPTER XXI.

Cells.—By the "efficiency" of a cell is meant the strength of current it will maintain through a given resistance, which is, of course, dependent on the E. M. F. and internal resistance of the cell, the rate of polarization and recovery, and also the "endurance" of the cell.

These measurements can be conveniently made in the following manner : the cell is joined up in series with a resistance, such as five ohms, and a key. Across the terminals of the cell is also joined a Weston voltmeter with low reading scale. The method is the same as that previously described for the measurement of battery resistance by fall of potential, and the connections are shown in Fig. 49. The voltmeter reading is first taken with the key open ; this gives the E. M. F. of the cell d_1. The key is then closed and the readings observed ; this gives the potential difference d_2 across the external resistance R. The internal resistance of the cell can then be calculated from the proportion,

$$d_2 : d_1 - d_2 :: R : X.$$

The cell is left on closed circuit, and the key opened just long enough to observe the E. M. F. at the end of every two minutes. It is then left on open circuit, and the voltmeter readings taken every two minute intervals, for ten minutes. From these data curves of the polarization and recovery can be constructed, using the times for ordinates and the E. M. F.s for abscissas. The "endurance" of the cell can be obtained by keeping a closed circuit through a known resistance until exhausted, and the ampere-hours or watt-hours calculated.

In place of a voltmeter, a galvanometer and high resistance, or a galvanometer and condenser can be employed.

Lamps.—The efficiency of a lamp is the ratio of the energy consumed to the candle power developed. It is calculated in watts per candle power. The connections for the measurement are shown in Fig. 97.

In series with the lamp is joined an ammeter and across the terminals a voltmeter.
The watts are obtained by multiplying the
volts by the amperes. The candle power is
observed by means of a photometer. If a
rheostat is placed in series with the lamp and

FIG. 97.

the resistance varied, the candle power and watts per candle
power at different voltages may be observed and a curve of
efficiency at these different voltages constructed.

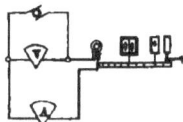

The "life" of the lamp is, of course, less the higher the
F. M. F. employed. This may be obtained for any given voltage
by leaving on closed circuit and observing the candle power
after different intervals of time. The efficiency gradually de-
teriorates, and after a certain time the candle power diminishes
to such an exent that it is no longer economical to use the lamp.

Motors.—The electrical energy given to a motor can be measured by joining a voltmeter across its terminals and an ammeter
in series with it. The electrical horse-power is then given by
the formula
$$E.\ H.\ P. = \frac{\text{volts} \times \text{amperes}}{746}.$$

If a Prony brake is employed, the mechanical horse-power
developed by the motor is given by the equation
$$M.\ H.\ P. = \frac{P \times S \times R \times 6.28}{33.000}$$

in which P = torque or pull in pounds, s = speed in revolutions
per minute, and R = the radius at which the pull is measured.

The efficiency is the ratio of the power developed to the
energy consumed; that is,
$$Efficiency = \frac{M.\ H.\ P.}{E.\ H.\ P.}$$

Transformers.—By means of the voltmeter and ammeter, the
energy given to a transformer can be measured, and in the
same manner the energy given out observed. The ratio of these
two values gives the efficiency.

Dynamos.—The commercial efficiency of a dynamo is the ratio
of the net output to the mechanical power applied to drive the
machine. The output, or the E. H. P., can be measured with the
voltmeter and ammeter, and the power consumed, or the M. H. P.,
by applying a Prony brake to the driving shaft. The efficiency
can then be obtained from the equation
$$Efficiency = \frac{E.\ H.\ P.}{M.\ H.\ P.}$$

The dynamo may also be run as a motor and the measurements made in the same manner as that given above;
$$Efficiency = \frac{M.\ H.\ P.}{E.\ H.\ P.}$$

CHAPTER XXII

MAGNETIC
DETERMINATIONS.

$$\begin{cases} Field \ (\mathcal{3C}) \\ Intensity\ of\ Magnetization\ (\mathfrak{I}) \\ Permeability \ \left(\mu = \dfrac{\mathfrak{B}}{\mathcal{3C}}\right) \\ Susceptibility \ \left(\dfrac{\mathfrak{I}}{\mathcal{3C}}\right) \\ Hysteresis. \\ Magneto\text{-}Motive\ Force. \\ Reluctance. \end{cases}$$

Field (𝒳).—The intensity of the magnetic force at any place, or the strength of a magnetic field, is the force which it exerts on a unit magnetic pole. The unit pole is defined as exerting on a similar pole at unit distance a unit force.

The C. G. S. unit of field density is the gauss, or one "line of force" per square centimetre.

The determination of the "horizontal intensity" of the earth's field, or any other very weak and uniform field, can be made by method of Gauss. The measurement depends on two observations, the time of oscillation of a magnet, and the angle of deflection caused by its action on another magnet. The first observation gives the product of the intensity of the field ($\mathcal{3C}$) and the magnetic moment of the magnet ($\mathfrak{M}$), or $A = \mathfrak{M}\,\mathcal{3C}$. The second gives the ratio of $\mathfrak{M}$ to $\mathcal{3C}$, or $\mathfrak{B} = \dfrac{\mathfrak{M}}{\mathcal{3C}}$. From these two results the value of either $\mathfrak{M}$ or $\mathcal{3C}$ can be found, thus :

$$\mathcal{3C} = \sqrt{\dfrac{A}{\mathfrak{B}}}.$$

The value of $\mathfrak{M}\,\mathcal{3C}$ is given by the equation :

$$\mathfrak{M}\,\mathcal{3C} = \dfrac{K\,\pi^2}{t^2\,(1 + \theta)},$$

in which t = time of a single oscillation of the magnet in seconds, K = moment of inertia of the magnet, θ = ratio of torsion of the suspending thread. If the magnet be of regular shape, the moment of inertia can be found by calculation from its weight and dimensions. The ratio of torsion of the suspending thread may be found by observing the deflection produced by twisting it through 360°, or, if this is small, it may be neglected.

The value of $\frac{\mathfrak{M}}{\mathfrak{K}}$ is obtained in the following manner : the large magnet, whose time of oscillation has been determined, is placed at a certain distance (r centimetres) from a magnetometer m, Fig. 98, and the tangent of the angle of deflection φ, obtained either by means of a telescope and scale, or a slider and sight moving directly on the scale.

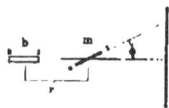

FIG. 98

For approximate work, the deflections of an ordinary compass needle can be taken in place of using a magnetometer. The magnet is then placed at a less distance, r', from m, and the angle of deflection φ^1 observed. From these observations the value of $\frac{\mathfrak{M}}{\mathfrak{K}}$ is obtained from the equation :

$$\frac{\mathfrak{M}}{\mathfrak{K}} = \frac{1}{2} \cdot \frac{r^3 \tan \varphi - r'^3 \tan \varphi^1}{r^3 - r'^3}.$$

If $\mathfrak{K}$ is accurately known in any given place, the field strength in any other place can be found by observing the time of oscillation of a magnet in the two positions, then :

$$\mathfrak{K} : \mathfrak{K}' :: t'^2 : t^2.$$

The value of $\mathfrak{K}$ can also be compared by suspending a magnet by a fine wire and determining the angle of torsion necessary to produce the same deflection in the two given fields.

Strong magnetic fields can be measured by Verdet's induction method. In this method, a small wire loop (of area f), connected with a galvanometer, is suddenly brought into or removed from the magnetic field, with its plane perpendicular to the lines of force, and the deflection (e) noted ; then $\mathfrak{K} = C \frac{e}{f}$, where c is a constant of the galvanometer.

The resistance of bismuth increases in a magnetic field and strong fields can be measured by a determination of this increased resistance and comparison with tabular values empirically determined.

Intensity of Magnetization ($\mathfrak{J}$).—This is given by the equation

$$\mathfrak{J} = \frac{\text{magnetic moment}}{\text{volume}}.$$

The value of the magnetic moment $\mathfrak{M}$ is obtained by the method given for $\mathfrak{K}$ from the equation

$$\mathfrak{M} = \sqrt{\mathfrak{M} \; \mathfrak{K} \times \frac{\mathfrak{M}}{\mathfrak{K}}}.$$

Permeability (μ).—The permeabilty of any magnetic material, such as iron, is the ratio of the magnetic flux ($\mathfrak{B}$) through the material to the field producing it—that is, $\mu = \dfrac{\mathfrak{B}}{\mathfrak{K}}$. The permeability of iron varies greatly according to the field strength, decreasing rapidly as it approaches saturation.

A convenient arrangement for the measurement of the permeability of small iron bars is shown in Fig. 99. Within a large rectangular piece of iron are placed the magnetizing coils s s'. The bars to be measured b b' are enclosed by the coils. A small coil, c, connected with a ballistic galvanometer, is held in position between the iron rods. When one of these is withdrawn, the coil c is thrown back by a spring, thus

FIG. 99.

cutting the lines of force of the field and producing a deflection of the galvanometer. Then if N be the total number of lines of force cut, or the total flux, $\mathfrak{B} = \dfrac{N}{A}$, where A is the area of the cross section of the bars in square centimetres. The value of N is found from the equation $N = K\ S$, in which $S =$ throw of galvanometer, and $K = $ constant. This constant can be determined by a method similar to that given for the absolute measurement of capacity and depends also on the number of turns and resistance of the exploring coil c. It can be determined by making use of a standard solenoid and calibrating coil.

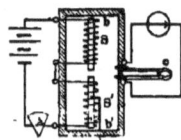

The value of $\mathfrak{K}$ is given by the equation

$$\mathfrak{K} = \frac{4 \pi n c}{10 l},$$

in which $n = $ number of turns in magnetizing coils, $c = $ current, in amperes, and $l = $ length in centimetres.

The permeability of rings of iron can be measured by a similar method. Upon the ring is wound a magnetizing coil, and also an exploring coil which is connected to a ballistic galvanometer. The deflection is then observed on either making, breaking, or reversing the magnetizing current.

The magnetometer can be used to measure the pole strength of long iron bars, when magnetized by a coil through which a known current is flowing, and the value of N found by multiplying by 4π.

Susceptibility (k).—This depends on the measurement of $\mathfrak{M}$ and $\mathfrak{K}$, and its value is given by the equation $k = \dfrac{\mathfrak{J}}{\mathfrak{K}}$.

Hysteresis, or magnetic lag, is conveniently observed by' the following method : two magnetizing coils s s', Fig. 100, are placed near a magnetometer and so arranged that no deflection is produced when a current is sent through them. The bar of iron, *b*, is then placed in one of the coils and the deflections noted with an increasing and decreasing current. From the values of these deflections and the strength of current used, hysteresis curves may be constructed.

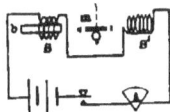

FIG. 100.

Magneto-Motive Force, or total magnetizing force, may be found for a solenoid from the equation :

$$\mathfrak{F} = \frac{4\pi N I}{10},$$

where $N =$ number of turns, and $I =$ current in amperes. The current is divided by 10 to reduce amperes to the c. g. s. unit of current. It is evident that the magneto-motive force = "ampere-turns " $\times$ 1.257 $\left(\text{the value of } \frac{4\pi}{10}\right)$. The practical unit is the ampere-turn.

Reluctance, or magnetic resistance, since it varies inversely as the permeability, also varies with the magnetizing force. Its value for a bar of iron is given by the equation :

$$\mathfrak{R} = \frac{l}{A\,\mu},$$

in which $l =$ length in centimetres, and $A =$ area cross-section in square centimetres.

The relations of the magnetic circuit are shown by the equation :

$$\text{Magnetic Flux} = \frac{\text{magneto-motive force}}{\text{reluctance}}$$

INDEX.

www.ingramcontent.com/pod-product-compliance
Lightning Source LLC
Chambersburg PA
CBHW021937190326
41519CB00009B/1048